# PARTIAL
# DISCHARGE BASED
# CONDITION MONITORING
# OF
# HIGH VOLTAGE EQUIPMENT

# PARTIAL DISCHARGE BASED CONDITION MONITORING OF HIGH VOLTAGE EQUIPMENT

**T. S. RAMU**
*Professor*
*Department of Electrical Engineering*
*Indian Institute of Science*
*Bangalore, India*

**H. N. NAGAMANI**
*Joint Director*
*Diagnostics, Cables & Capacitors Division*
*Central Power Research Institute*
*Bangalore, India*

PUBLISHING FOR ONE WORLD

**NEW AGE INTERNATIONAL (P) LIMITED, PUBLISHERS**
New Delhi • Bangalore • Chennai • Cochin • Guwahati • Hyderabad
Jalandhar • Kolkata • Lucknow • Mumbai • Ranchi
Visit us at www.newagepublishers.com

Published by New Age International (P) Ltd., Publishers
First Edition: 2010

---



***Branches:***

- 36, Malikarjuna Temple Street, Opp. ICWA, Basavanagudi, **Bangalore**-560 004. Tel.: (080) 26677815, Telefax: 26615247, E-mail: bangalore@newagepublishers.com
- 26, Damodaran Street, T. Nagar, **Chennai**-600 017. Tel.: (044) 24353401, Telefax: 24351463, E-mail: chennai@newagepublishers.com
- CE-39/1016, Carrier Station Road, Ernakulam South, **Cochin**-682 016. Tel.: (0484) 2377004, Telefax: 4051303, E-mail: cochin@newagepublishers.com
- Hemsen Complex, Mohd. Shah Road, Paltan Bazar, Near Starline Hotel, **Guwahati**-781 008. Tel.: (0361) 2513881 Telefax: 2543669, E-mail: guwahati@newagepublishers.com
- No. 105, 1st Floor, Madhiray Kaveri Tower, 3-2-19, Azam Jahi Road, Nimboliadda, **Hyderabad**-500 027, Tel.: (040) 24652456, Telefax: 24652457, E-mail: hyderabad@newagepublishers.com
- RDB Chambers (Formerly Lotus Cinema) 106A, Ist Floor, S.N. Banerjee Road, **Kolkata**-700 014, Tel.: (033) 22273773, Telefax: 22275247, E-mail: kolkata@newagepublishers.com
- 16-A, Jopling Road, **Lucknow**-226 001. Tel.: (0522) 2209578, 4045297, Telefax: 2204098, E-mail: lucknow@newagepublishers.com
- 142C, Victor House, Ground Floor, N.M. Joshi Marg, Lower Parel, **Mumbai**-400 013. Tel.: (022) 24927869, Telefax: 24915415, E-mail: mumbai@newagepublishers.com
- 22, Golden House, Daryaganj, **New Delhi**-110 002. Tel.: (011) 23262370, 23262368, Telefax: 43551305, E-mail: sales@newagepublishers.com

---

**ISBN : 978-81-224-3092-9**

**C-10-02-4306**

Printed in India at Glorious Printers, Delhi.
Typeset at In-house, Delhi.

**PUBLISHING FOR ONE WORLD**

**NEW AGE INTERNATIONAL (P) LIMITED, PUBLISHERS**
4835/24, Ansari Road, Daryaganj, New Delhi - 110002
Visit us at **www.newagepublishers.com**

# Preface

Power supply companies, national and regional power generation and transmission authorities spend large sum of money every year to maintain high reliability of supply to users. The reliability of the whole system is a strong function of the reliability of each component of an electricity generation and transmission system, for example, the Generator, the Transformer, the Circuit breaker, the Cables and the like. Often, power equipment may be composed of sub-components, for example, a transformer is made in two distinct parts, the electromagnetic section and the bushing section and hence the long term performance of these parts depends on different aspects. Major attention is paid in the Book to the electrical insulation structure, but other components are also given due consideration.

The design of equipment insulation for long service life based on deterministic design considerations is not possible, since the accompanying failure mechanisms are always stochastic. In view of this, physical as well as statistical models are required to understand the processes leading to insulation degradation. The origin of this subject dates back to the beginning of this century. The starting point of the physical models for insulation failure was on a phenomenological basis. Chemists of that time, Eyring, Arrhenius, and later on Dakin, laid the foundations for treating the failure as a natural consequence of a slow physico-chemical process, called Ageing. The term "Ageing" of insulation has somewhat similar connotation as applied to ageing of human body.

Ageing is always thought to occur under a stimulus, such as electric, thermal, mechanical and environmental stresses and, for the ageing to be initiated, a certain minimum degree of stressing, called the threshold stress, is necessary. However, probabilistically speaking, this inevitable process starts at any value of stress, however small. Two aspects of interest here are the time that elapses before failure ensues, or the magnitude of stress/ stresses that can be applied for a desired service life. In either case, a probabilistic approach is indicated, because even under nominally identical conditions, these quantities are associated with a high degree of scatter.

The term 'failure' is subjective in the sense that a product is rarely run to complete breakdown, rather it is taken off the system when it shows signs of malfunction, either in the quality of power delivered or by forcing a reduction in the amount of power drawn from the system. When so qualified, it becomes incumbent on an insulation designer to seek the opinion of the power supply utilities as to what, in their decision, the 'end-of-life' (or the end point) criterion is. The determination of the optimal operating stresses and the estimation of the corresponding life is based on two important considerations, the heuristic modeling and model validation. As has been said earlier, there are two features in modeling- the physical (deterministic) and the stochastic (probabilistic). In the recent past, the probabilistic approach has gained particular prominence, at least in engineering science. This is due to the fact that no material either insulating or conducting used in the technology applications, is ever produced in its pure state and nor it is practically feasible to do so. The statistical models are intensely data driven, in the sense that a large amount of credible experimental data collected on the equipment in question is required. Now the question comes on what kind of experiments are to be conducted, destructive (for example, voltage breakdown in which the equipment is subjected to such stresses so as to cause a failure), or, non-destructive (diagnostic).

The diagnostic parameters considered essential are many. The insulation enclosing defects of various kinds behaves in a different way than does a defect free material. For example a monolithic structure in the body of which are present dielectric defects, such as, a gaseous inclusion of arbitrary shape, a gaseous cavity as it is often called, breaksdown electrically under certain condition of operation. The breakdown is initially confined to a small region and as the process continues, gives rise to a small electrical discharge, called the partial discharge (PD). Depending upon the number, the shape and the pressure of the gas in the cavity, measurable magnitudes of energy is liberated. This PD energy slowly erodes the insulation and is a known cause of ageing and premature failure of insulation. It has been observed that the PD and several of its sub-parameters are effective diagnostic indices for predicting failure, or estimating the life of power equipment in service. The physical processes leading to PD are more or less well established, however, the way the PD quantities are to be considered to understand ageing of insulation is still a matter of considerable debate.

A few important contributions have come on this topic, by Kruger, Bartnikas (edited book) and others. The intention of the authors is to summarize the contents of these books, add some recent material generated by them and make the information on Partial Discharges accessible to a larger section of the readers including practicing engineers in utilities so as to encourage them to use PD parameters as possible indices of insulation degradation. Attempts have been made to include practical data and case studies on actual equipment at site. The authors are open to constructive suggestions and criticisms for improving the quality and content of the book.

The Chapter on Acoustic Technique for the detection of partial discharges and all discussions thereof, has been contributed by Dr. Mrs. Nagamani.

The Authors are grateful to many of their colleagues who have supported them on various counts, carrying out laboratory and in-situ experiments, supplying the PD data. Particular mention is made here to our distinguished teacher and most revered mentor, Prof. B. I. Gururaj for his unstinted support, critical and constructive comments and encouragement in writing this book. We would have been highly gratified had Prof. Gururaj seen the final version of the book. We offer this book as a humble tribute to him. One of the authors, Mrs. Nagamani, wishes to thank her colleague Mr. T. Bhavani Shanker (Engineering Officer, CPRI), for providing useful field data and information on onsite measurements.

The Authors express their heart felt thanks to the members of their families for their constant cooperation and support during the preparation of the manuscript.

Authors also wish to thank M/s. New Age International Publishers, New Delhi, India and in particular, Sri. S.P. Sudarshan, Assistant Manager, Bangalore office.

***The Authors***

***Bangalore, India***

# Contents

CHAPTER

1

# Introduction to PD

## 1.1 GENERAL

Power equipment employ insulation to isolate energized conductors at different potentials and high voltage conductors from ground. Electrical isolation is also necessitated for reasons other than purely electrical purposes. In power apparatus, insulation may also be called upon to perform auxiliary functions such as cooling and internal electrical discharge suppression purposes. It is important to note that all the three states of matter are involved in the process of electrical isolation of conducting parts. Over the past five decades and until now, insulating materials of different origin have been employed. In electrical equipment design among others, atmospheric air, inorganic gases like sulfur hexafluoride ($SF_6$), nitrogen and organic gases such as carbon dioxide are used in some applications. Plastic materials, find applications in a majority of electrical equipment. Fiberglass or mica reinforced epoxy resins are among insulating materials of choice in rotating machines. Organic polymers such as polyethylene derivatives, ethylene propylene, rubber and similar class of materials are predominantly used in power cables up to the highest voltage. Natural polymers such as mineral oil, cellulose and its derivatives have long been used as liquid insulated systems in power transformers.

## 1.2 DEFECTS IN INSULATION – MICRO AND MACRO

A ***'defect'*** may be defined broadly as a species of unintended material included in the matrix of insulation. The presence of defect has a tendency of adversely influencing the bulk properties of the chosen insulating material. The size of defects in insulation will be generally in sub-micron level whereas insulation thickness will be in the range of millimeter. Depending on the size, defects are classified as micro and macro defects. Some of the examples of micro defects are, air bubbles (or gaseous packets), voids, micro-cracks, conducting particles in $SF_6$ system, sharp edge of electrodes in capacitors and the like. Some of the macro defects are, semi conducting particles in extruded cable insulation, sharp projection of semi-conducting layer into cable insulation, improper contact between insulation and conductor surface, poor stress cone used in cable joints and end terminations, physical absence of semi conducting layer over cable insulation, de-lamination of varnish in winding wire, inter-turn fault in transformer winding, etc.

The process of building or formation of insulation structure involves several stages starting from selection and preparation of raw materials, process of application of raw materials, thermal or chemical treatment when found essential, provision of protective enclosures, and so on. The steps involved are simpler in low voltage apparatus, for example, conductors of overhead distribution lines have ambient air as main electrical insulation with supporting insulators at poles/towers. Whereas, it may involve very complex procedure of extrusion and cross linking of polyethylene at high pressure and temperature and then cooling to ambient temperature with huge manufacturing plants as in the case of high voltage Cross Linked Polyethylene (XLPE) cables. The total process of providing electrical insulation involves men, materials and machines and also different environmental conditions.

It is therefore, practically not possible to achieve a perfect electrical insulation without defects as it may get contaminated and disturbed during material handling, processing and manufacturing of electrical equipment. Defects may also develop in electrical apparatus during its operation under the influence of thermal, electrical, mechanical and environmental stresses imposed on it. Defects present in new apparatus may lead to accelerated ageing of insulation and premature failure of apparatus.

Defects in high voltage power equipment like power transformers, machines, switchgears, underground cables, capacitors and gas insulated systems are more critical as they lead to electrical discharges. The most common types of defects in the majority of the electrical equipment are (*a*) defective or improper material, design and technique employed, (*b*) voids, imperfections, impurities, foreign particles, moisture in insulation, (*c*) stress concentration at the electrode edge, (*d*) poor workmanship (*e*) inadequate technique to prevent ingress of moisture during manufacturing, transportation, storage, installation or in service, (*f*) deterioration of insulation under the influence of thermal, electrical and mechanical stresses, (*g*) exposure to transient over-voltage due to switching and lightning impulse. In addition, there could be defects specific to equipment, which may lead to electrical discharges. Defects in some of the major equipment are listed in Appendix 1.

The defects which do not result in internal electrical discharges beyond the safe limits are harmless to the insulation. Insulation design in power equipment has to take into account the possible entry of non-substantiating components and precautions to be taken to safeguard the intended components.

## 1.3 PARTIAL DISCHARGE (PD)

An electrical breakdown of a medium has been defined as an irreversible process in which a medium under consideration is unable to support the applied external electric field. This definition is not strictly applicable for an insulation system. The breakdown therefore, shall be re-defined keeping in view the components thereof. The effect of the relatively weaker dielectric (e.g. a gas) or a defect can be detailed as under. In Fig. 1.1(a) the expressions for capacitances $C_a$, $C_s$ and the electric fields $E_a$, $E_s$ with usual notations can be written as:

$$\left(\frac{E_a}{E_s}\right) = \left(\frac{C_a}{C_s}\right) \tag{1.1}$$

It is seen from Eqn. (1.1) that the air medium is stressed more than the solid and since the electric strength of air is much less than that of the solid medium it is possible that the air space breaks down electrically, an electric spark of considerable energy spans the air medium. However, the system itself does not breakdown. Since the spark or electrical discharge as it is called, is restricted only to the weaker component, it is called a partial discharge (PD). In an exactly similar way, unintentionally introduced gaseous media either in series or in parallel with dielectrics, are of higher permittivity as shown in Fig. 1.1(b), would cause internal discharges in the air films. Extending this process further, in Fig. 1.1(c) an air filled cavity situated in the matrix of insulation is also electrically stressed far in excess of the surrounding medium and a partial breakdown initiates in the cavity. There is a substantial difference between the breakdown of gaseous cavity and the thin air space shown in Fig. 1.1(b), for example, a film of air bounded between two non-equipotentials, the dielectric slabs behave electrostatically in a much different way than a film bounded between a dielectric surface and an equipotential, a metal electrode. This aspect is considered in detail later.

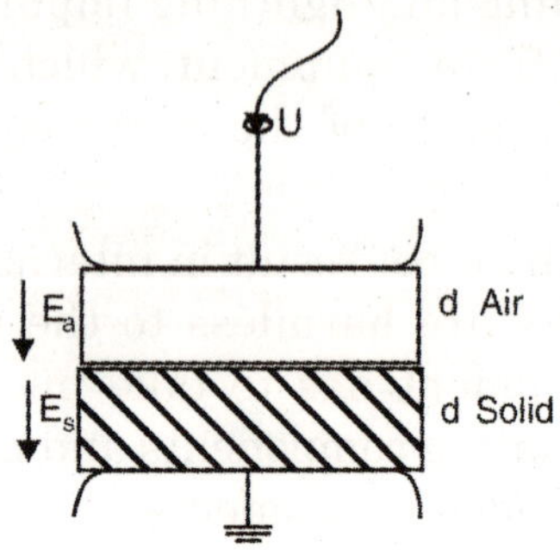


**Fig. 1.1(a).** Air gap and solid insulation

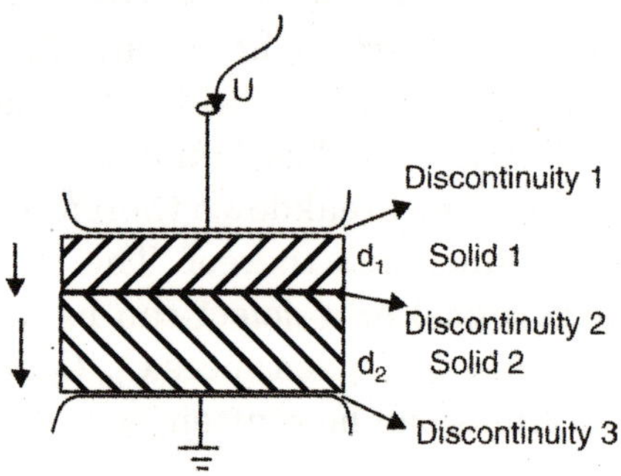


**Fig. 1.1(b).** Two different solid insulation media

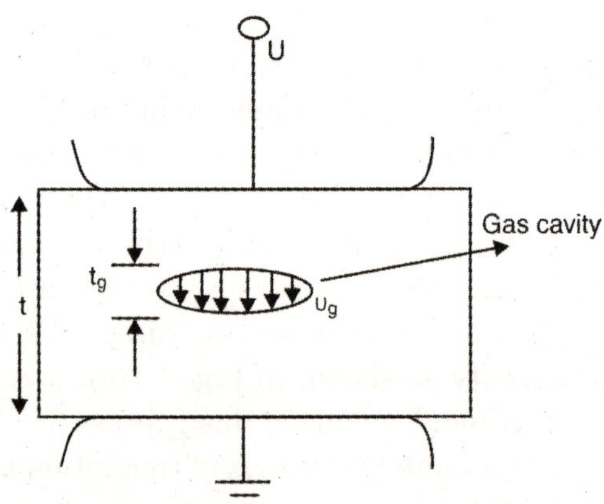


**Fig. 1.1(c).** Gas cavity in solid insulation

It is important to note that single insulating material is generally not used in power equipment (except in, extruded power cables). Concomitantly, a system concept is indicated. Behaviour of a system, although intimately connected with the components thereof, is found to be essentially different when the stresses to which they are subjected are of different origin. Avoiding details, the system approach for various operating reasons are very much more involved than when single insulating structure is involved. To highlight this aspect, consider an insulation system consisting of two different solid components as shown in Fig. 1.1(b). It can be seen that in addition to two solid phases, there are three interfaces or boundaries making the system discontinuous in the direction of applied electric field. The discontinuities in this arrangement are thin air spaces (which are unintentional) which are forced to support the electric field in excess of their nominal withstand levels. This causes a local electrical field enhancement and hence a local breakdown. In this sense, regions such as these, introduced for whatever reasons, constitute weak sites in which charge generation and annihilation process constituting an electrical discharge is initiated. In the process of integration of different insulating phases to make up the final insulating structure in an equipment, weak links are introduced and the behaviour of the system depends more on the weak links than on the substantiating components, $d_1$ and $d_2$.

When a dielectric continuum does not involve a liquid medium, disruptive electrical discharges of various origins are known to occur. Even in liquid impregnated system, PD is possible at highly stressed locations. In order to ensure quality performance over long times, of electrical insulation in power equipment, it is imperative that a substantial consideration needs to be provided to eliminate local incipient breakdowns. Given that the PD mechanism is very closely associated with minute, highly (electrically) stressed regions as might be occurring around conductors having sharp or pointed geometrical shapes, equipment design and manufacturing process shall eliminate conductor corners or sharpness to avoid the initiation of PD.

From this discussion, it is clear that two aspects together decide the possibility of occurrence of PD

i. Presence of conducting regions with very small radii.
ii. Defects' location, shape, nature, number density and their distribution in the matrix of insulation will decide whether or not PD occurs under any given condition.

Depending on the location of defect and electrode geometry, partial discharges from the defects are classified as (*a*) internal discharges occurring in the defects contained within the insulation (*b*) surface discharge occurring along the surface of the solid insulation and (*c*) corona discharges due to concentration of high stress at sharp electrode edge.

## 1.4 PD MECHANISM

The processes leading to partial breakdown of gaseous cavities surrounded fully by dielectric is a function of the instantaneous electrical charges of either polarity, present on the surface of the solid. The formation of electrical charges is due to the difference in the polarizability between the gaseous and the solid medium. Every individual discharge is associated with energy of its own. Also, the magnitudes of energies of seemingly identical discharges are known to be different. The need for studying physics of partial discharges and development of techniques for quantification and measurement of discharges arises because of the fact that the energetic charges bombarding the surface of the gas filled cavity have a tendency to interact with the chemical structure of the insulating materials. The result of such interactions is the breakage of chemical bonds and subsequent erosion of the material in question. Also, it has been reported that a considerable local heating confined to small regions in the body of insulation takes place adding further to the degradation process. Admittedly, the degradation due to PD is a slow process and eventual failure of insulation due to this effect might sometimes be overtaken by other processes. From the stand point of insulation design it is essential to recognize that PD is known to be a precursor to other rapidly developing degradation failure mechanisms.

The quantification of partial discharge process is, at this point of time, a subject of considerable debate. The charge, the energy, the repetition rate and other similar quantities are often used to quantify partial discharges.

## 1.5 PARTIAL DISCHARGE AS AN INDEX OF INSULATION DEGRADATION

Partial discharges have long been used as a diagnostic tool for identifying internal defects in power equipment. It has also been used in the estimation of elapsed or remaining life, albeit with little success. It is of particular advantage to employ this parameter since it can be sensed at terminals of an equipment. When measurements are carefully performed, and ensuing data analyzed efficiently, by a knowledgeable person, PD quantities would serve as a reasonable diagnostic tool in monitoring the status of electrical insulation in power apparatus. The PD diagnosis is of particular relevance in 'Dry-type' insulation for the reason that the most predominant mode of failure of this type of insulation is due to internal discharges in the gaseous defects often enclosed in the body of such insulation. Moreover, being a terminal measurement it is a non-invasive method to track the slowly changing fault initiation and propagation.

In equipment specific diagnosis, a matter of great concern is to identify the most sensitive and accurately measurable quantity from among those mentioned earlier. Often, a function of the PD quantities which has a direct bearing on the long time behaviour of power equipment, is employed to detect any developing faults or the introduction of new fault types into the system. A consideration is given to the PD pulse description such as phase resolved, magnitude resolved and number density resolved techniques. These, the so called PD signature analysis or pattern recognition methods, have come into vogue some time ago. These modern techniques add quality information to PD diagnosis of power equipment.

Partial discharge patterns sometimes reveal the nature and severity of the defect present. For example, if gas or air filled cavities, or when fissures and cracks are present in the insulation matrix, the PD patterns show marked difference in the shape and angular occupancy of the discharges on the input waveforms. Also, parasitic internal and external noise can sometimes be distinguished from the real PD pulses.

## 1.6 CORONA DISCHARGE

Partial electric discharges from a conductor at high voltage exposed to the air medium, such as transmission line conductors, transformer bushing terminals, sharp protrusions, edges and corners of high voltage conductors and the like, are the sites from which electrical discharges called as Corona discharges occur. Corona discharges are often the source of considerable parasitic noise and every attempt shall be made to eliminate, or, minimize them. Besides creating difficulties in measurement, corona gives out interference signals in the radio frequency range. Corona noise couples with communication signals and corrupt them. International standardizing bodies have established maximum permissible magnitudes of the corona noise in terms of what is called Radio Interference Voltage (RIV). Electric power stations have to mandatorily abide by these guidelines. The corona from a single conducting protrusion produce unmistakable signature and is easy to detect. However, if the number of these protrusions is large, as is the case of transmission lines and line accessories, the noise pollution in the ambient becomes inordinately large and equipment PD measurements are untenable. This is particularly more serious in site measurements.

## 1.7 DEFECT CENTRIC PD PROCESS

Much has been said about the partial breakdown in gaseous cavities in insulation and the ensuing PD. It is possible that other defects, such as conducting and dielectric defects, other than gaseous, may also be present in an insulating structure. For example, a pointed high voltage electrode seeing an electrical field at its tip far in excess of the applied field around the tip due to stress concentration may be thought of as a defect. The Maxwell stress due to electrostatic forces at this point is several times greater than the strength of the material, and hence, a local breakdown occurs which in turn opens up a fissure of microscopic dimension. This fissure has all the features of a gas filled defect in the insulation matrix and from hereon the classical PD process takes over. In this way, whether defects are of the conducting or of the dielectric type, PD can be triggered in a similar way. It is now clear that the PD process is defect centric, that is, it is initiated in and around a defect site in an otherwise continuous phase of insulation.

## 1.8 PD INSTRUMENTATION

The techniques used in the detection and measurement of PD also depend on the quantities used to quantify the discharges. The more important measuring techniques designed to detect PD are based on the effect produced by PD itself. A large body of literature to suggest that PD activity is associated with several electrical, thermal, physical and chemical effects such as electromagnetic transients, dielectric losses, mechanical or acoustic (pressure) waves, sound, light and chemical reactions. It is therefore, possible to use these quantities as detection parameters in addition to charge and energy measurements.

In all measurements, similar to and including the ones mentioned above, the external environment and the measuring equipment itself introduce electrical noise. The origin and the magnitude of the noise, its frequency spectrum, should be carefully explored. The interference caused by the noise to the PD signal is perhaps the most important aspect of PD measurements. It has often been seen that the actual PD pulses are buried in excessive noise due to the ambient and external electrical circuit in the vicinity of the test object. Elimination of noise of various kinds is a difficult task. In the recent past powerful electronic and computer aided techniques have been developed to try and eliminate or, at least minimize the noise. The power source, such as the measuring transformer used in the PD measurement also introduces electrical noise. However, modern power sources are carefully designed so as to be free from incipient discharges.

Depending upon the test object, its terminal capacitance and operating voltage level, a situation may arise wherein the demand on the capacity of the power source might become inordinately large. For example, in testing long lengths of power cables and the stator insulation of large generators this situation is normally encountered. There are methods of mitigating this situation as shall be explained later in this book.

Any measuring equipment should adequately address issues relating to the sensitivity and resolution of the quantities being measured. In addition, the frequency spectral characterization of the discharge train and associated time varying quantities are to be considered in some detail. This introduces further constraints and complications in the selection of a class of measuring equipment so as to be able to produce repeatable and reproducible measurements.

An elementary block diagram of a general partial discharge detection system based on the most widely used conventional electrical method is shown in Fig.1.2. The more important components, besides high voltage high frequency filters, thereof, are the following:

i. PD free voltage source

ii. Detector

iii. Coupling capacitor to provide a impedance path for the discharge signals emanating from the test object and divert them to the measuring impedance

iv. Input transformer and measuring impedance
v. Pulse shaping network
vi. Shielded cables
vii. Data acquisition/display/recording device
viii. Calibrator

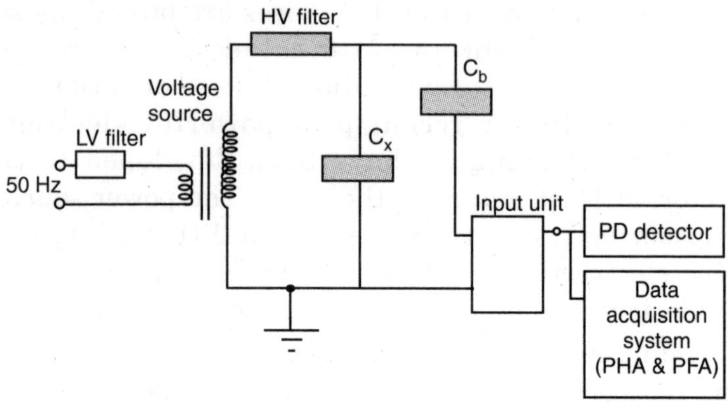


**Fig. 1.2.** Block diagram of a general partial discharge detection system

Partial discharge detection in DC systems is much more complex to understand. Avoiding details at this juncture, discharges under direct voltage are in the form of direct current pulses. The measurements are performed in markedly different ways compared to AC case, as described later.

## 1.9 PARTIAL DISCHARGE DATA ACQUISITION AND ANALYSIS

Partial discharge is a random process in which electrical charge transfer is highly chaotic in space and time. This means that the number density of discharges, as well as their magnitudes, is distributed in a statistical sense. Standard procedure for data analysis is not generally possible and requires the invocation of more involved probabilistic methods. The data analysis part is perhaps the most vital aspect of a partial discharge measurement. The analytical procedure, starting from setting up of the PD measurement experiment, data acquisition, signal recognition and feature extraction, to the inference assimilation, often requires the services of an experienced experimenter, or an expert system, carefully designed to effectively replace him. The later aspect is given particular attention in modern PD data acquisition and analysis.

An aspect of great importance in PD measurements is that the actual signal being measured is often shrouded in a maze of parasitic interferences of various origins. Detection, extraction and validation of signals have been found to be difficult in view of the largely differing PD corrupted signature in the insulating structure of different power equipment. Simpler frequency filtering techniques are untenable, because, it often happens that the Fourier decomposed frequency spectra of the signal and noise coalesce thus rendering the filtering difficult. In fact, the PD measurements are more involved with the noise elimination protocols than the measurement itself. It is therefore felt pertinent to dwell considerably into this topic. A significant section on the subject of measurement has been devoted to detail techniques for noise suppression, signal extraction and validation.

## 1.10 NON-CONVENTIONAL TECHNIQUES FOR PD MEASUREMENT

The conventional electrical PD detection method is the most widely employed method for almost all types of electrical equipment like, cables, transformers, gas-insulated systems (GIS) and is generally as per IEC 60270. Electrical PD measurement has certain limitations particularly, when applied to test objects with large capacitance and for on-site testing. New test techniques such as electromagnetic (VHF and UHF) and acoustic methods have come into vogue in the recent past and are classified as non-conventional. These are relatively new techniques and are gaining importance as they can be applied on-line for on-site PD diagnosis. With great advancement in data acquisition and computational techniques, sensing technique, the non-conventional methods are gaining greater acceptance. More details on these techniques are covered in separate chapters.

## 1.11 CALIBRATION

Before commencing a partial discharge measurement, it is necessary to ascertain whether the measured magnitude of the discharge is properly referenced against a known standard charge pulse. The purpose of this exercise is of two folds:

i. To rectify the charge magnitude detected at the terminals of the test object against the standard.
ii. To estimate, if any, the attenuation of the signal magnitude measured between any two desired locations.

The estimation of the circuit attenuation is essential in PD measurements, because, it is often possible that one would be measuring a magnitude much less than the real value due to loss of signal amplitude. Also, signal distortion normally takes place in the presence of reactive circuit components either inside the test object or in the path leading to the detector. Low voltage lead-in cables also introduce distortion due to harmonics. The process of calibration is carried out as often as is desired, but should certainly be undertaken when a test object is changed or the associated components are replaced. Standard methods for calibration of the detector and the test circuit are available for use in PD measurements.

## 1.12 CONDITION MONITORING BASED ON PD

Temporal changes in the condition of insulation in certain power equipment and the amount of damage accumulated thereof is often estimated using the partial discharge quantities. There is some disagreement on how much credence one can place on the assessment based on PD and more so the parameters used in such an endeavour. Avoiding details, the general tendency is to apply energy related quantities, for example, the PD energy per cycle of the power frequency wave, the number density of pulses in each half cycle and their phase expanse, incremental power loss factor referenced with the applied voltage and the like.

Be that as it may, a reasonably accurate judgment of the status of insulation can only be made using several properties of insulation which are known to be sensitive to stress related ageing including PD. If necessary, an intelligent weighting function may be attached to each of the parameter considered based on sound premises, which is hard to come by unless the analysis is carried out by an experienced person with a good knowledge of the nuances of PD related ageing in insulation. This action calls for careful design of experiments on prototype models of the equipment considered including the stresses to which it might possibly be called to take-on in service. Long duration accelerated stress tests are indicated, but the PD measurements themselves being made at, for example, stresses at or slightly above the normal operating levels. The book provides substantial information on these topics with suggestions to design stress profiles to be applied and inspection protocols for different equipment used in power systems.

It is also relevant to mention here, that, secondary effects produced by high energy internal discharges might have a more pronounced propensity for accelerating the ageing in insulation than the PD itself. The book describes these mechanisms and provides approximate methods for estimating the enhanced degradation.

CHAPTER

2

# Processes Leading to PD

## 2.1 GENERAL

Partial discharge belong to a kind of gas discharge occurring in solid dielectric systems. Upon application of an electric field, the gas molecules trapped in gas filled cavities (defects), in extruded solid systems, are ionized by impact collisions of electrons releasing localized low intensity, low energy free charges. The impact of collision interactions of these charges (with non-zero momentum) with the walls of the cavities cause high energy densities confined to microscopic sites. If this energy is greater than the bond energy of the continuum, local erosion or pitting takes place.

In organic polymers, besides this effect carbonization of the walls of the cavity can also result. It may be seen that there is a distinct possibility that the cavity becomes nearly conducting, in time, due to the deposition of virgin carbon. This secondary formation often manifests as a conducting inclusion in the otherwise insulating structure. Although this process is highly confined, depending upon the number density of the defect sites in a unit volume of the material, it gives rise, among other things, to enhanced conductivity of the structure. Several research workers [1, 2] have observed photoemission and acoustic emission during such a process. Thermal effects, in general, are disregarded as the temperature rise in the material due to these effects are generally small.

As has been said earlier, in practice, no insulation is used in its virgin state and combination of several insulating materials for a particular application makes the design and production more difficult. Besides this, the mechanism of insulation degradation is material and stress specific, requires one to invoke a system concept. Also for various applications, different kinds of insulation are used in combination. During the process of manufacture electrical insulation encloses a large number of defects.

The defects are broadly classified as conducting defects and dielectric defects. The Fig. 2.1 illustrates the possible defects in insulation and their location. Depending upon the type of defect two distinct nature of failure can be visualized. When the defect is conducting, the process of failure is called 'Treeing'. On the other hand dielectric inclusions like gaseous cavities give rise to internal electrical discharges, called as partial discharges.

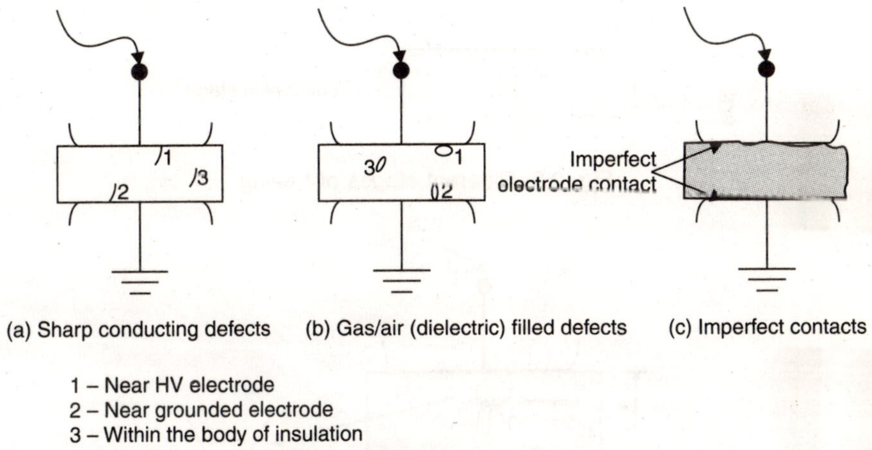


**Fig. 2.1.** Breakdown due to defects

## 2.2 PARTIAL DISCHARGES DURING TREEING

Treeing is a process of breakdown due to pointed conducting (metallic) or semi-conducting defects. Most often, treeing occurs in monolithic or extruded or cast polymeric materials. The initial process of treeing, called the inception, begins at the tip of the conducting point shown in Fig. 2.2 by way of a minute pre-breakdown crack or fissure-like formation. This kind of treeing is called electrical treeing. If there is a discontinuity in the body of the dielectric which is filled with a polar liquid or an electrolyte like water or its solutes the tree formation is called electro-chemical treeing. Both types of trees progress and later cause a disruptive breakdown of the dielectric in two distinct stages after inception as shown in the Fig. 2.3.

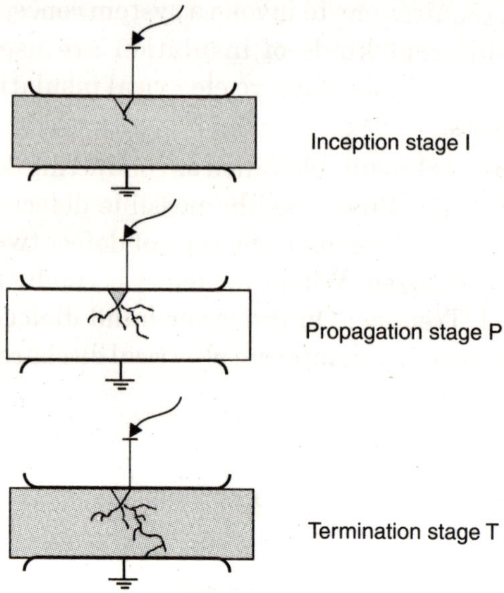


**Fig. 2.2.** Different stages of treeing

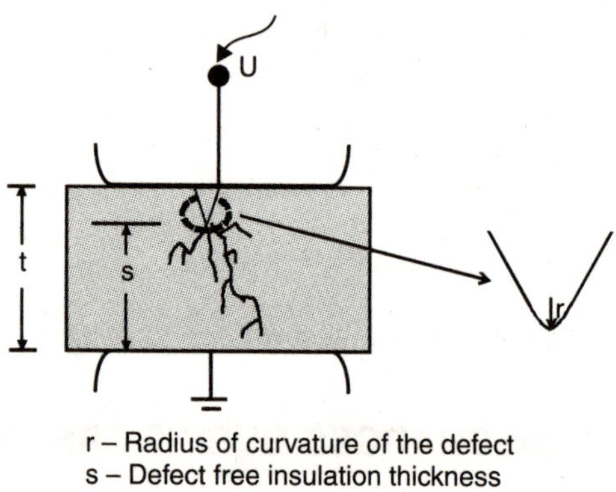


**Fig. 2.3.** Effect of the tip radius of a metallic protrusion

The electrical treeing also occurs at the site of conducting or gas filled defects in a thick dielectric. The name 'electrical' is used here to indicate that the processes are due only to the electric field in association with a defect not involving an electrolyte.

Inception is caused by an extremely high electric field at the tip, Fig. 2.3. The maximum alternating electric field at the tip of a sharp defect in a solid insulation is given by the Eqn. 2.1.

$$E_{max} = \frac{2U}{r \cdot \ln\left(1 + \frac{4S}{r}\right)} \tag{2.1}$$

The tree progresses from the time of inception in two further stages, propagation and termination.

The field at the tip $E_{max}$ is very much higher than the normal electric field ($U/t$) depending upon the radius of the tip. At this high electric field, the dielectric experiences an electrostatic force which when exceeds the disruptive mechanical force required to cause a crack, the first mechanical breakdown channel ensues. The potential at the tip of the defect is conveyed to the tip of the crack so formed in two ways;

(*i*) In organic insulation the extremely high electrostatic force due to $E_{max}$ tends to dissociate the structure and during this process virgin carbon is known to be a byproduct. These carbon particles are slowly deposited on the tree channel rendering it nearly conducting. The crack tip now becomes a virtual extension of the defect from which a subsequent crack in a random direction would occur.

(*ii*) In inorganic insulation, the extension of the crack tip is not immediately possible since the inception channel does not become conducting in the way explained above. One of the theories extended to explain further progression of the tree proposes that partial discharges are initiated in the tree channel. The energy associated with PD may be sufficient to erode the material and preferentially around the region of the crack tip. However, there are objections to this theory in that the pressure in the tree channel is very low and the conditions for triggering PD are very rarely met.

Consider a gas filled cavity of microscopic dimensions, included in the volume of a solid insulating slab shown in Fig. 2.4. The voltage across the cavity, electrically, is a function of the enhanced electric field, the value of which is given by $E \times (\varepsilon_{r\text{solid}} / \varepsilon_{r\text{cavity}})$. As the applied voltage is increased, the gas cavity breaks down at $U_g$ corresponding to the minimum of the Paschen's graph shown in the Fig. 2.5. This means that the breakdown field is not only a function of $U_g$ and the thickness $t_g$, but also of the pressure of the gas in the cavity. This means that the breakdown voltage of the cavity of a known thickness has a series of values depending upon the gas pressure. Depending upon the structure of the tree, essentially two different formations are observed in practice, tree-like tree and bush-type tree (Fig. 2.6). A tree-like tree forms when the tip of the conducting defect is very sharp. The subsequent development of the channel is very fast and termination occurs in such a fashion that a fine channel bridges the tip of the high voltage electrode and ground. A bush type tree on the other hand occupies a larger volume

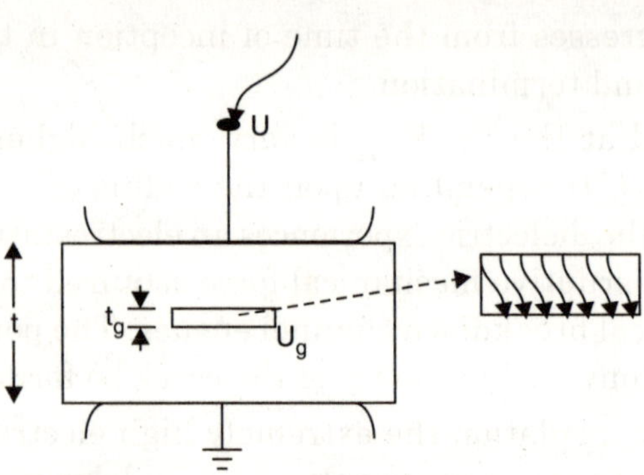


**Fig. 2.4.** Breakdown of a gas cavity enclosed in an insulating slab

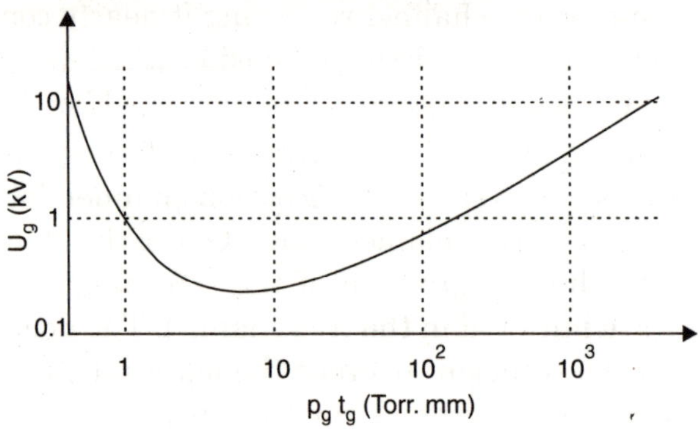


**Fig. 2.5.** Paschen's Law for a narrow, gas filled cavity types of tree

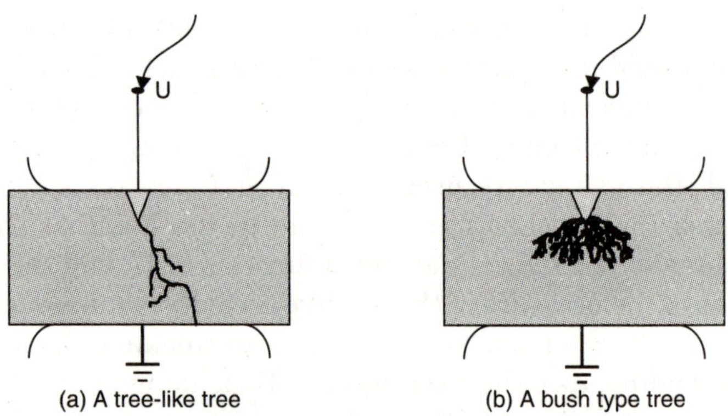


**Fig. 2.6.** Electrical tree formations

of the dielectric. This type of tree is formed when the tip of the defect is blunt. Sometimes a bush also forms when a gaseous phase is in the immediate vicinity of the defect. A large volume of the dielectric is destroyed by the bush, however, the time taken for breakdown is much longer.

## 2.3 CORONA AND PARTIAL DISCHARGES

As has already been mentioned, insulation encloses various kinds of defects during manufacturing, processing, storage and transport. Gas filled defects surrounding a high voltage conductor, gas cavities, in the body of insulation and improper electrode-insulation interface are the major causes of disruptive electrical activities. Charge transport within such cavities give rise to a pre-breakdown process in which the cavity electrically breaks down and the enclosed charge is transferred on to the immediate surroundings of the cavity in the direction of the applied field. The partial discharges as they are called are localized within a small volume into which electrical energy is pumped in, causing local breakage of the chemical bonds of which the insulation is made. The process leading to PD around a conductor in free space is called Corona discharge.

### 2.3.1 Corona Discharges

Consider a spherical conductor of radius $a$ carrying a charge $Q$. Referring to Fig. 2.7, the potential, $\Phi$, at any point $r$ from the center of the sphere in space is given by:

$$\Phi = \frac{Q}{4\pi\varepsilon_0 r} \tag{2.2}$$

The corresponding electric field, $E$, Eqn. (2.3), is the directional derivative of $\Phi$ and radially outward/inward depending upon the polarity of the charge.

$$E = \frac{Q}{4\pi\varepsilon_0 r^2} \tag{2.3}$$

Recognizing that, $4\pi\varepsilon_0 aU = Q$ the expression for $E$ becomes,

$$E = \frac{aU}{r^2} \tag{2.4}$$

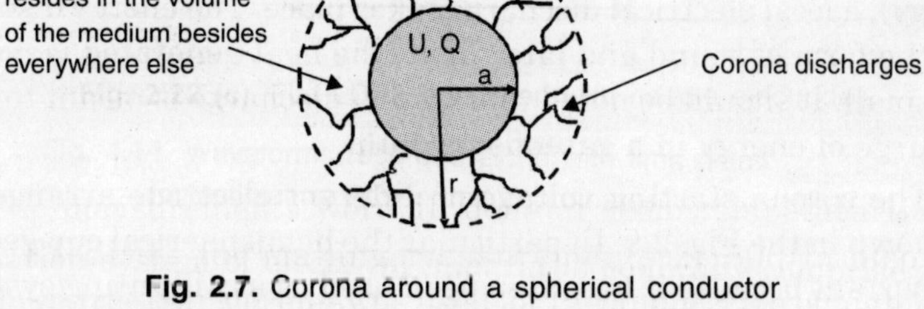


Fig. 2.7. Corona around a spherical conductor

The immediate vicinity of a charged sphere, is therefore, electrically stressed to the extent of the energy residing in a volume element around that region given by $2\pi\varepsilon_0 aE^2$. If this electric field is greater than the breakdown strength of free space, corona discharges emanate from the sphere.

A similar process will take place around a conducting rod (Cylinder), shown in Fig. 2.8. A conducting rod in space charged to a potential $U$, behaves as a discontinuity for the electric field at its extremities ($A$, $B$ in the Fig. 2.8). The electric field in the vicinity of $A$, $B$ is much higher, (depending upon the field enhancement factor, $\xi$) compared to the region $C$. In particular the regions surrounding rims at $A$ and $B$ are highly stressed and causes intense ionization (Chemical Dissociation) of the surrounding gaseous medium.

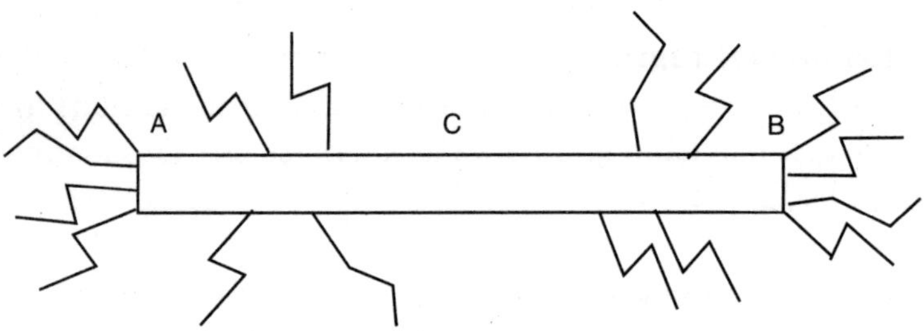


**Fig. 2.8.** Corona from a cylindrical conductor

Two important aspects to be noted here are:

i. If the extremities are sharp-edged, then the ionization field is 1% to 0.5% of the corresponding field around a sphere of equivalent radius.

ii. The lengths of corona discharge paths are much longer than for a corresponding sphere.

The corona discharges associated with the other conductor geometries in free space can be visualized in a similar way. The mechanisms leading to electrical corona discharges are summarized below. If the energy density in the surrounding medium exceeds the ionization potential (energy), a local electrical discharge takes place. The energy discharged is in the form of sound and later light, the heat generated is normally very small. It should be noted that corona discharge is meant to denote discharge of energy in a gaseous medium.

The corona starting voltage of different electrode arrangements are shown in the Fig. 2.9. In particular the hemispherical cup geometry is important because it has often been used as a calibrating system for PD measurements [3].

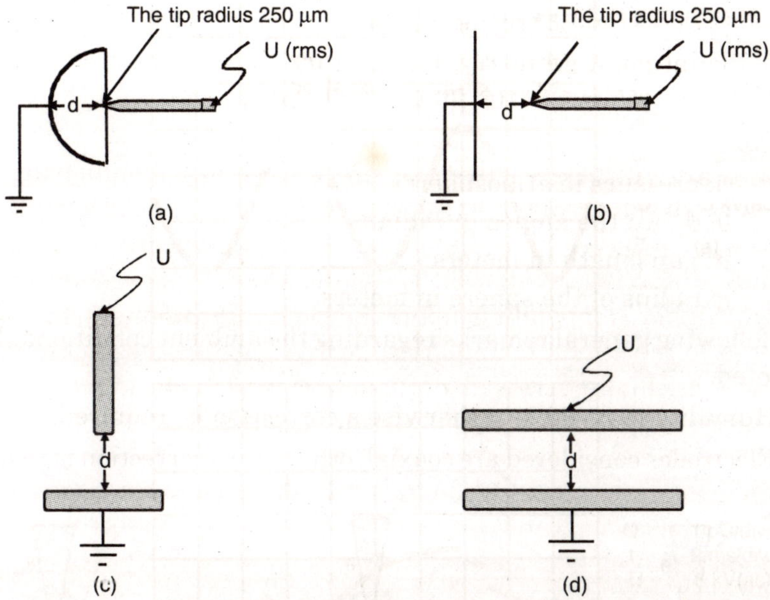


**Fig. 2.9.** Gap geometries for which corona starting voltages given in Fig. 2.10

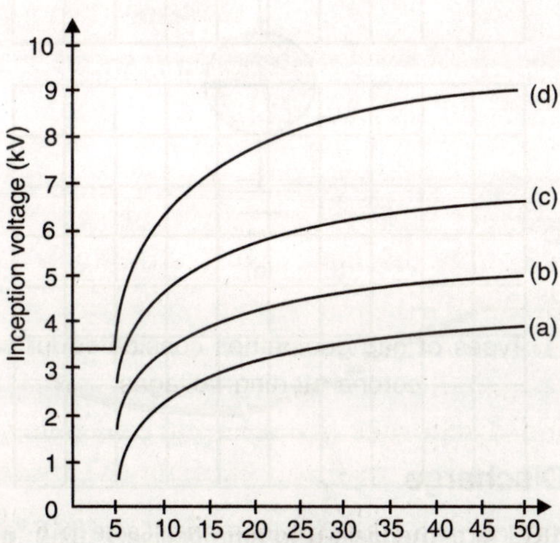


**Fig. 2.10.** Corona starting voltage for different gap geometries

Empirical expressions for calculating corona starting (Inception) voltages ($U_i$), of some important gap geometries are given below [4].

i. Point-plane geometry, Fig. 2.11(*a*)

$$U_i = 1.08*10^5 \ (\theta^{3.5} d^{1.35} p^{1.25}) \tag{2.5}$$

ii. Point-sphere geometry, Fig. 2.11(*b*)

$$U_i = 35*10^5\ (\theta^{3.2}d^{1.3}p^{1.5}) \quad (2.6)$$

iii. Point-point geometry, Fig. 2.11 (*c*)

$$U_i = 1.6*10^5\ [(\theta_1\theta_2)^{2.25}d^{1.35}p^{1.5}] \quad (2.7)$$

where;

$p$: pressure in atmospheres
$\theta$, $\theta_1$, $\theta_2$: the angles in radians
$d$: gap length in meters
$r$: radius of the sphere in meters

The following general remarks regarding the ambient conditions should be noted:

- Humidity max. 60 % , otherwise a correction is required.
- Electrodes considered are coaxial, otherwise a correction is required.

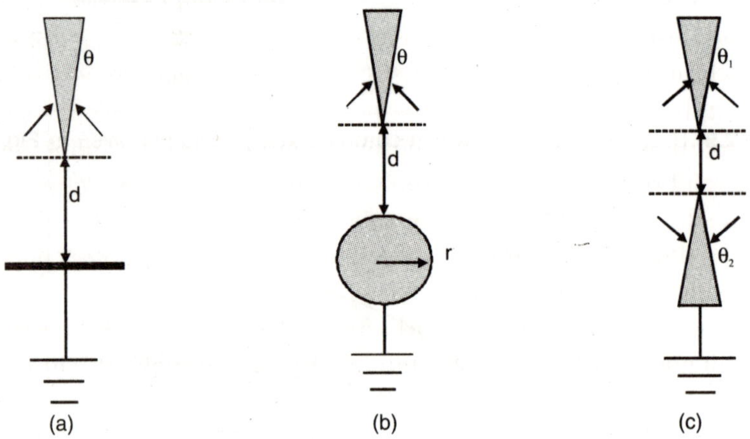


**Fig. 2.11.** Types of gap geometries considered for calculating corona starting voltages

### 2.3.2 Partial Discharge

It is possible that corona-like process can take place in the body of a solid or a liquid media where gaseous inclusions (Gas Cavities) exist. In such an event the discharges are notionally called as partial discharges. The term 'Partial' is meant to convey the meaning that the discharges do not span the thickness of the insulation. Prior to the occurrence of PD, the walls of the cavity form an interface with the dielectric where Maxwell-Wagner polarization comes into effect. The interfacial charges supplement the enhanced field due to the cavity. If the voltage across the cavity due to this field exceeds the Paschen minimum, a discharge takes place by which the cavity charges are neutralized. The process of neutralization ignites a local discharge of electricity, the partial discharge. The time required for the process is much shorter (<<1μs) than the duration of the periodic applied voltage. These processes release a considerable (measurable) energy into the insulation matrix in a very short time and hence become a cause for degradation of insulation. A large volume of insulation is thus rapidly destroyed. If the degree of imperfection is large, the number of discharge sites and hence the total energy in PD per Hz is correspondingly very large.

## 2.4 MECHANISM OF PD

The occurrence of electrical charge accumulation and transport process in dielectric structures enclosing inhomogeneities has been known for quite some time now. Scavenging of charges of either polarity deposited on the walls of an arbitrary shaped gaseous inclusion due to an applied external electric field takes place due to the collapse of the dielectric strength of the cavity. This, the partial breakdown as it is called is due to the applied electric field and the component of the electric field in its direction caused by the charges on the walls of the cavity, called the cavity field. As shall be explained later, this mechanism depends on several parameters for example the shape and size of the cavity, the properties of the gas in the cavity including the pressure. There are certain conditions to be fulfilled for the first partial breakdown to occur and for completely neutralizing the charges. The breakdown gives rise to a fast rising current pulse (or charge pulse) which can be sensed at the terminals of the object in question. It has been observed that there is a strong dependence of the voltage waveform on the breakdown mechanism. In the following, these effects are described in some detail.

### 2.4.1 Partial Discharges under Alternating Voltage Conditions

For the sake of simplicity, consider a plane parallel specimen of insulation to which a time varying (sinusoidal) electric field is applied. The process of charging and recharging can be understood using an equivalent circuit approach as shown in Fig. 2.12.

If $C_0$ is the capacitance of the (single) defect site, $C_1$ and $C_2$ are the unbroken capacitances of the specimen and $U_i$ is the Paschen minimum, the gas cavity breaks down at $U_i$, called the discharge inception voltage, as shown in the Fig. 2.13. In Fig. 2.13(a), the closing of a fictitious

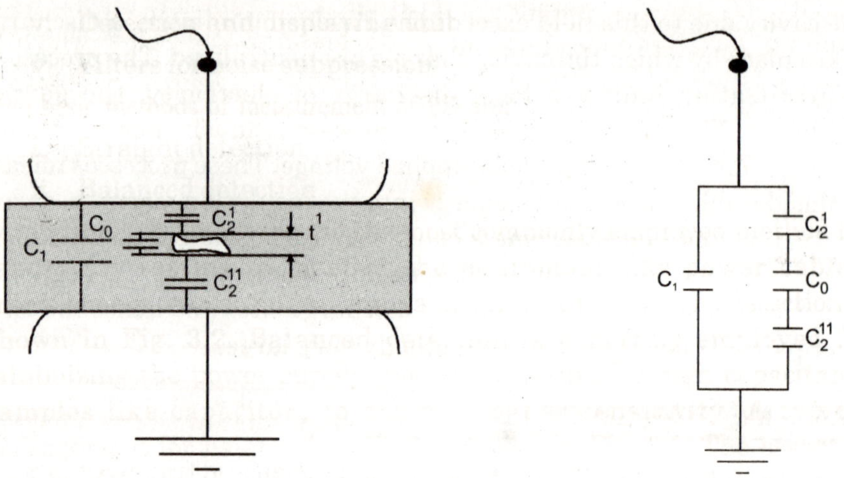


(a) (b)

**Fig. 2.12. (a)** Gaseous defect in a solid dielectric; **(b)** Equivalent circuit of (a)

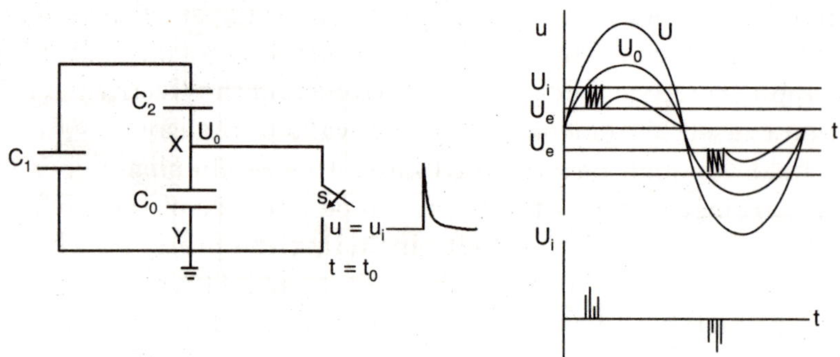


(a) (b)

**Fig. 2.13. (a)** General PD pulse; **(b)** PD pulse on the voltage wave form

switch represents neutralization of the charges on the walls of the cavity of capacitance $C_0$. The instantaneous charge transferred from a point of higher potential (indicated X) to the one at lower potential (indicated Y), results in a very fast rising transient voltage pulse, Fig. 2.13(b).

The spark generated due to the instantaneous annihilation of charges in the cavity, neutralizes a certain amount of the charge residing on the walls of the cavity. The resulting displacement current (or the charge pulse) can be observed at the terminals of the specimen.

When the neutralization is nearly complete, the cavity acquires a potential, $U_e$ called the extinction voltage. The potential $U_i$ is then is exponentially restored upon further interfacial charging. The magnitude of the charge pulse, $q_0$ , depends on the geometry of the cavity, $U_i$ and the voltage $U_e$. Further it is seen that the $U_e$ is the voltage level beyond which the spark endures and below which the cavity is devoid of surface charges. Effectively, $U_i - U_e = \Delta U$, is the voltage pulse due to the annihilation of the charges.

The following are the approximate expressions for PD charge transfer.

$$\Rightarrow q_0 = C_0 \Delta U$$

$$\Rightarrow q = C_2 \Delta U \tag{2.8}$$

$$q_0 = \frac{q(C_0 + C_2)}{C_2} = q\left(\frac{C_0}{C_1} + 1\right) \approx q \tag{2.9}$$

In the above expressions, $q_0$ is the actual cavity charge. Obviously it is not possible to measure $q_0$ directly. The charge, $q$, the apparent charge, flowing out of the terminals of the specimen is a measurable quantity and it can be shown that $q \propto q_0$.

The energy, $E$, associated with the discharge of magnitude, $q_0$ , present in a cavity with capacitance $C_0$, is responsible for causing irreversible structural damage to the insulation and this can be computed as follows:

$$E \cong \left(\frac{1}{2}\right) C_0 \Delta U \cdot U_i = \left(\frac{1}{2}\right) C_0 U_i q_0 \tag{2.10}$$

$$E = \left(\frac{1}{2}\right) C_0 \left[U_i^2 - U_e^2\right] = \left(\frac{1}{2}\right) C_0 [U_i + U_e] \Delta U \tag{2.11}$$

$$= \left(\frac{1}{2}\right) C_0 \Delta U \cdot U_i \left[1 + \frac{U_e}{U_i}\right]$$

$$\Rightarrow E \cong \left(\frac{1}{2}\right) U_i \cdot q$$

Also, the charges in the cavity in the post-spark regime would result in an increased electrical conductivity in the dielectric matrix due to pseudo conducting inclusions, the charged cavities and hence a loss of energy.

#### 2.4.1.1 Interfacial PD Effects in Heterogeneous Systems

In a structure composed of a weaker dielectric medium, like a gas, and a stronger medium like a solid, the PD processes at the interfaces have been found to be interestingly different than has been understood until now [5]. Referring to Fig. 2.14(a), the media mentioned above have been bounded by equipotentials (conducting planes). An increase in the applied voltage causes PD to occur in the gaseous medium which increases the mean dissipation factor measured at the terminals up to a maximum value of the applied voltage and decreases sharply thereafter as shown in Fig. 2.15(a). This process can be explained by the fact that when the air gap completely breaks down, the dissipation factor is due to dielectric effects only and the partial discharges do not seem to contribute to dissipation factor.

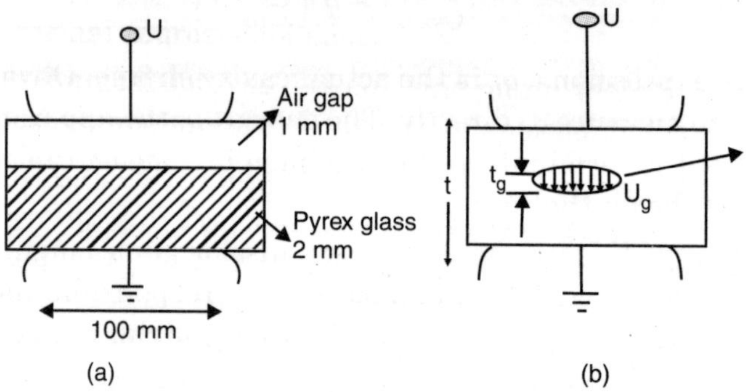


**Fig. 2.14.** Dielectric arrangements for measuring energy loss

Supposing the air inclusion is in the body of the insulation and the insulation is bounded by the equipotentials, Fig. 2.14(b), the dissipation factor does not show a discontinuity of the type shown in the Fig. 2.15(a). In such an arrangement, one would expect that the dissipation factor to continuously increase and remain stationary beyond a certain voltage level as shown in Fig. 2.15(b). This shows the effect of the position of the gaseous defect in the matrix of the insulation and its proximity to the electrode.

The energy loss per unit volume does not seem to include these effects as shown in Fig 2.15(c). It seems possible, that, after the incidence of breakdown of the gas medium, increase in energy due to this effect is not translated in real terms because PD energy is much less than the energy due to the dielectric losses of the solid medium.

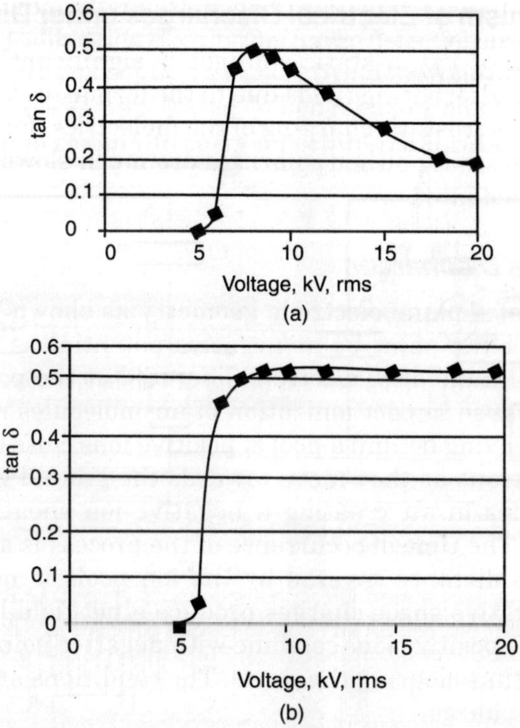


**Fig. 2.15.** Dissipation factor vs stress for the arrangements shown in Figs. 2.14(a) & (b), respectively

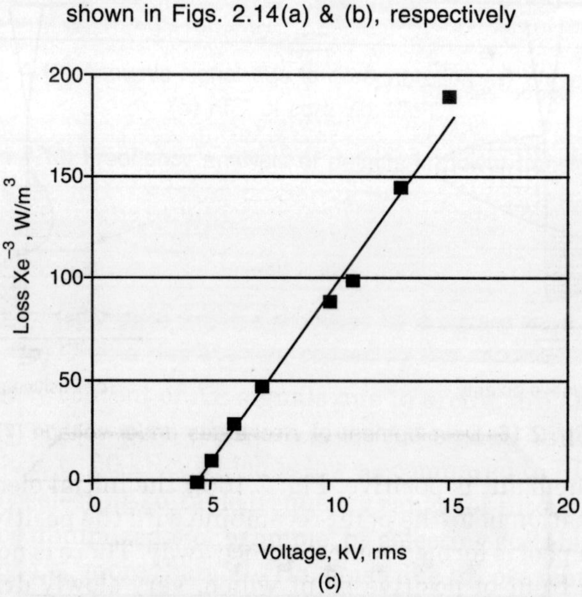


**Fig. 2.15 (c).** Energy loss due to discharges in the arrangement shown in Fig. 2.14(a)

### 2.4.2 Mechanism of Electrical Discharges under Direct Voltages

The mechanism of PD in static voltages is significantly different [6] from that under alternating fields due to the formative time effects and mechanism of electrostatic charging of the dielectrics. Furthermore, the discharges in air under either polarities are much slower and fewer as shall be explained later.

#### *2.4.2.1 Corona Discharges*

Consider a point – plane electrode geometry as shown in Fig. 2.16(a) and 2.16(b). Let the point be at negative potential as shown in Fig. 2.16(a). A small number of electrons liberated at the point are highly energetic and cause impact ionization of air molecules releasing more electrons and leaving behind a pool of positive ions close to the negative point. The electrons as they move towards the ground get attached to oxygen molecules in air creating a negative ion sheath at a certain distance away. The time of occurrence of the process is almost 1 nano - second and the distance covered by the ion pools is nearly equal to $10^{-4}$m. The negative space charges produce a net field lower than the initial value, the positive ions combine with negative point and negative ions to ground thus clearing the space. The conditions are now feasible for a second discharge.

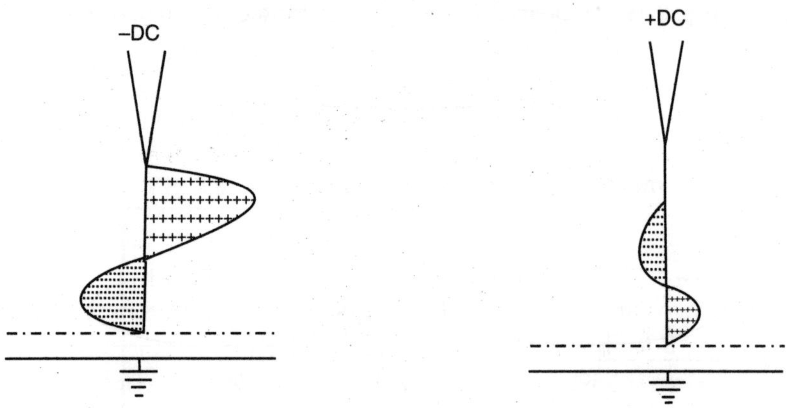


(a) Negative polarity (b) Positive polarity

**Fig. 2.16**. Development of discharges under voltage [7]

When the point is positive, Fig. 2.16(b) the initial electrons, after causing ionization near the point recombine with the positive point and a positive ion pool is formed some distance away. There is now a positive ion shield in front of positive point which, very slowly drifts away to ground and soon thereafter a second discharge ensues. Figure 2.17 shows the recurrence of discharges under AC and negative DC voltages.

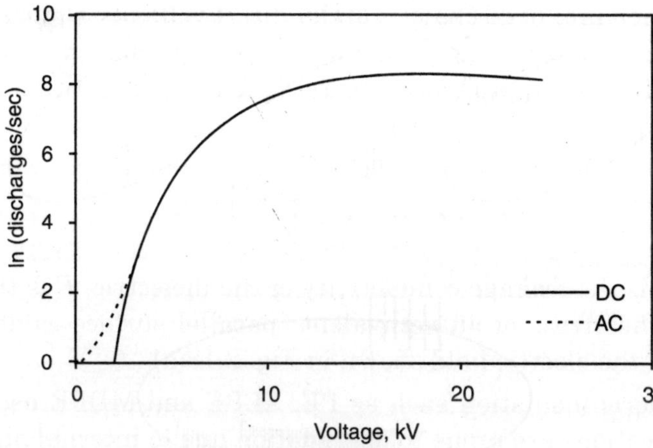


**Fig. 2.17.** Frequency of discharges under AC and negative DC voltages

### *2.4.2.2 Partial Discharges in Solid Insulation*

Under the application of direct voltages, the equivalent circuit of a solid dielectric including a gaseous cavity is to be modified as shown in Fig. 2.18. The voltage distribution in the insulation body is predominantly due to the leakage resistance (dissipative element). However, the capacitance of the cavity, its surface conductance and the capacitance of the unbroken dielectric are introduced to give effect to the charge storage and dissipation in the system.

The charge and hence the voltage appearing across the cavity is a function of the conductance $g_1$ and capacitance $C_1$ of the dielectric in series with $C_C$. In view of a continuous leakage of charge, the cavity stress stabilizes over by a period of time and at the Paschen minimum condition the void is cleared of the cavity charge which appears as a "current pulse" at the terminals.

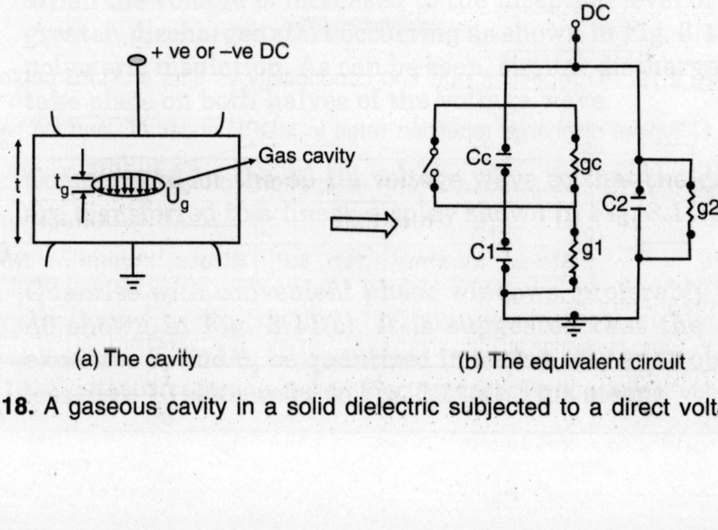


**Fig. 2.18.** A gaseous cavity in a solid dielectric subjected to a direct voltage [8]

The repetition rate of discharges under direct voltages is much smaller than that under alternating voltage and the energy density of discharges is also smaller. An approximate expression for the repetition rate, *f*, is given by [7]:

$$f = 1.13 \times 1011\ \sigma \times \left(\frac{E}{E_p}\right) (\mathrm{s}^{-1}) \tag{2.13}$$

in which σ is the leakage conductivity of the dielectric, $E$ is the stress and $E_p$ is the stress of an equivalent parallel surface cavity in the direction of the electric field shown in Fig. 2.18(a).

Polymeric insulation such as PE, XLPE and MDPE used in DC cable applications are prone to degradation due to internal discharges of this type. A laboratory model cable specimen subjected to a direct voltage was observed to give out a repetition rate shown in Fig. 2.19.

A comparison of typical discharge repetition rates in MDPE under alternating and direct voltages is provided in Table 2.1.

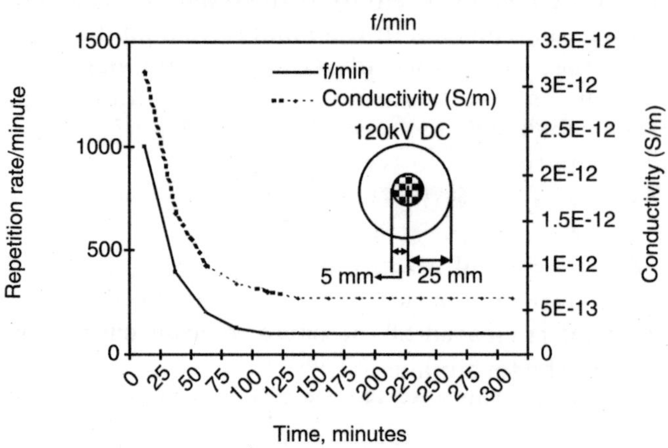


**Fig. 2.19.** PD repetition rate and conductivity vs time in XLPE cable

**Table 2.1.** Typical discharge repetition rates in MDPE under AC and DC voltages

| Cavity | Form of applied voltage | | | |
|---|---|---|---|---|
| | Alternate | | Direct | |
| | Stress, kV/mm | N/h, $10^5$ | Stress, kV/mm | N/h, $10^5$ |
| Spherical | 6 | 9 | 60 | 4 |
| | | | 120 | 8 |
| Cylindrical | 4 | 9 | 40 | 4 |
| | | | 80 | 8 |

## 2.5 DISCHARGE MAGNITUDE AND ENERGY

Referring to Fig. 2.18, the discharge magnitude $q$ can be calculated by knowing the capacitances

$$q = \left(C_2 + \frac{C_1\,C_c}{C_1 + C_c}\right) \times \left(\frac{C_1}{C_1 + C_c}\right) \times U_i \tag{2.14}$$

where $U_i$ is the voltage across the cavity at inception. It should be observed that in the case of DC, the equivalent circuit also involves the incorporation of conductances. In view of this the actual magnitude of discharge $q$, in terms of applied voltage $U$ can be written as,

$$q = \left(C_2 + \frac{C_1\,C_c}{C_1 + C_c}\right) \times \left(\frac{C_1}{C_1 + C_c}\right) \times \left(\frac{g_c}{g_b + g_c}\right) U \tag{2.15}$$

The energy loss in watt-sec, $W$ is

$$W = \frac{1}{2}\left(C_2 + \frac{C_1\,C_c}{C_1 + C_c}\right) \times \left(\frac{g_c}{g_b + g_c}\right)^2 \times V_i^2 \tag{2.16}$$

The effect of electric stress on the discharge-repetition rate is shown in Fig. 2.20. At any particular stress, the discharge-repetition rate decreases with time, the effect becoming less pronounced at the higher stresses. Cavities of large diameter yielded higher discharge repetition rates indicating more than one discharge site per cavity. The discharge-repetition rate and the volume conductivity is lower and the times to breakdown longer for specimens made with paper washed in deionized water rather than normal filter water.

On lowering the temperature from 293 K to 90 K, the discharge-repetition rate and discharge magnitudes for a cavity within oil-impregnated paper decreases as shown in Fig. 2.21. The decrease in the repetition rate is expected as the volume conductivity is also reduced at lower temperatures.

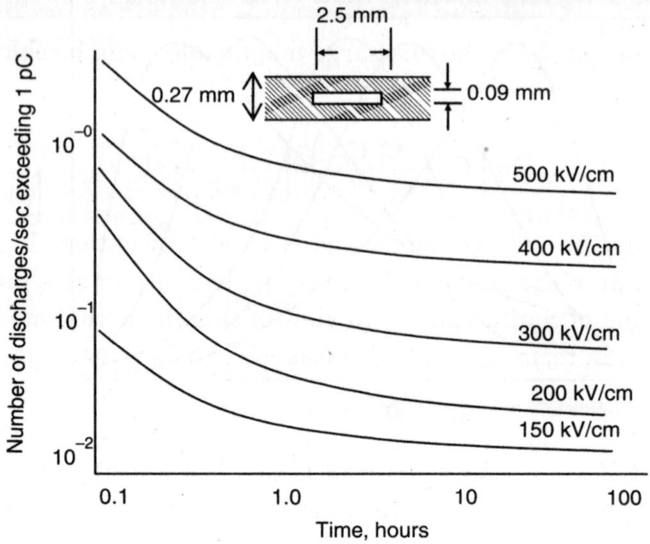


**Fig. 2.20.** Variation of discharge repetition rate with time for oil impregnated paper [9].

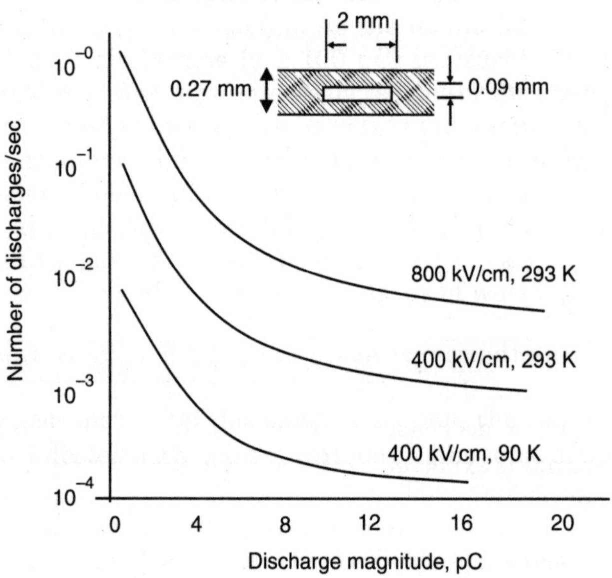


**Fig. 2.21.** Discharge magnitude and repetition rate in oil impregnated paper [20]

## 2.6 PULSE-LESS OR MICRO DISCHARGES

Prior to and shortly after the inception of a discharge in the highly restricted cavity, field aided charge generation due to pinch effect seems possible. It follows that the charge annihilation is neither instantaneous nor pulsive. However, a finite and nonzero leakage current, not due to the known pulsive PD ensues at the terminals. This effect has been called Pulse-less PD or micro or µ-discharges or silent discharges. This process was recognized, among others, by Lemke, Karner, Konig, Ramu and Smith [10 – 13]. The distinguishing features of these discharges are:

- they are very minute, less than a third of the smallest measurable pulsive PD
- essentially a very low frequency phenomenon, (a few Hz to a maximum of a few kHz)

It is thought that the µ- discharges are responsible for insulation ageing in a very different but definite way by which treeing and similar electrostatic processes leading to ageing take place.

As opposed to partial discharges, the pulse-less discharges appear more continuously. Actually, the µ-discharges contribute to a greater energy loss in insulation at relatively low voltages. It is therefore imperative that a study of these effects on a physical basis and to establish tangible correlations between insulation ageing and µ discharge parameters is of great importance in the design high reliability electrical insulating systems. Since µ-discharges are essentially low frequency processes and hence, detection containing wide band or tuned amplifiers cannot be used for signal identification. Special, low noise, high gain amplifiers with very low distortion are to be used. This means, conventional partial discharge measuring apparatus cannot be used. Also, in view of the very small charges involved, it is neither possible, nor practicable to measure the individual discharges. With this in view, an integrating type discharge measuring circuit should be thought of.

Elaborate arrangements have to be made to realize noise-free measuring environment to the extent possible. First of all, a darkly shielded cage with a primary shielding kiosk with a second internal shielding box both made of non-magnetic material has to be installed. These shielding points are to be grounded to separate earth and are segregated from the earth used for the measuring circuits. This arrangement reduces, and in some cases eliminates both radiative stray signals and mains borne signals.

In the early development of the bridge at KEMA, one of the balancing points (Point B in Fig. 3.17(a)) of the bridge was connected to the chassis for operational simplicity and the lower point of the bridge was earthed through an insulating support. Later, however, it was realized that this arrangement gives rise to considerable unwanted radiative coupling with the chassis acting as an antenna. Another drawback is that the chassis acting as an antenna and that it is at high voltage. To get over this problem, the common mode point AB is now separate from the chassis, which is then grounded. This has reduced the noise, and considerably facilitated the operation of the bridge. To further reduce the harmonics in the supply voltage, a capacitive filter (harmonic suppressor) Fig. 3.17(b) has been used.

The level of voltage that can safely be allowed to appear across the capacitor $C_1$ has to be much higher than in a conventional bridge circuit (where it is about 1 volt) for improving the bridge sensitivity. In the integrating bridge developed here, this voltage is between 300-500V. This means the bridge lower arms are usually at this voltage. It is therefore essential, from the standpoint of operator safety, the measuring points must be isolated (by using electro/optical devices) from the detector (oscilloscope). In the present set up, electro/optical (E/O) and opto/electrical (O/E) conversion devices linked by a fiber optic cable have been used.

Experiments under laboratory conditions have been performed at the KEMA laboratories, the Netherlands to understand the process of pulse-less PD [10-13]. Specimens used in the series of tests are shown in Fig. 2.22(a) and Fig. 2.22(b), respectively called as CIGRE type I

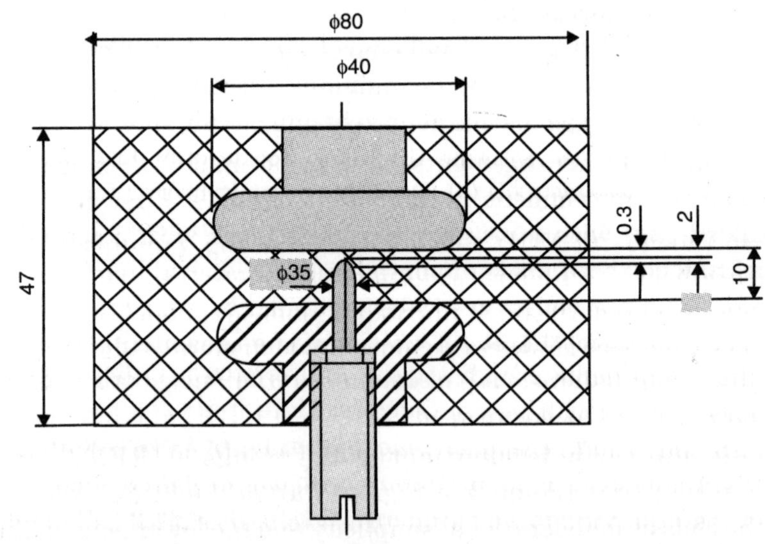


**Fig. 2.22(a).** CIGRE-I and II Round Robin epoxy specimen arrangement

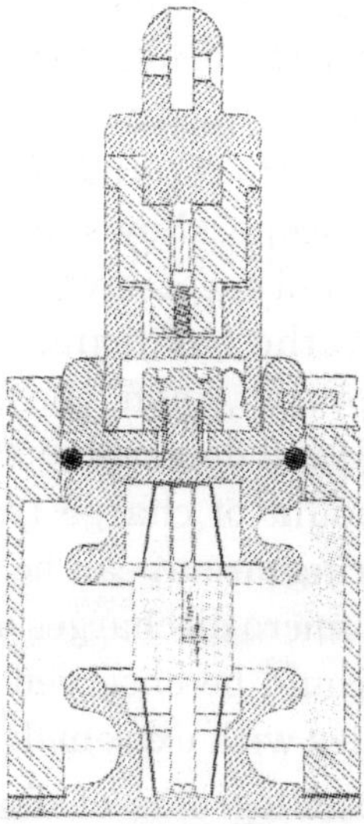

**Fig. 2.22(b).** Filled and unfilled epoxy specimens with needle electrode (tip radius 5μm and 1mm airspace)

and II cells. The dielectric in question is unfilled epoxy resin with deliberately included gaseous defects of known geometry. Kodall [14] describes defect geometries such as air filled hemispherical cavity. Figure 2.22(c) is a plain polyethylene (LDPE) slab with a needle high voltage electrode against a flat low voltage electrode. It has been observed that using this geometry μ discharges could be initiated at relatively lower voltages and easily resolvable pulse shapes could be obtained.

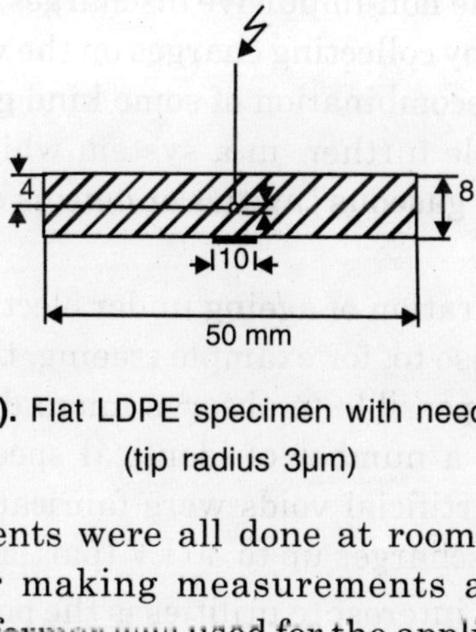


**Fig. 2.22 (c).** Flat LDPE specimen with needle electrode (tip radius 3μm)

The measurements were all done at room temperature and at 50Hz. Facilities for making measurements at 500Hz also exist. Discharge free transformer was used for the application of high voltage to specimens. The specimens, transformer and measuring circuits (discharge measuring bridge) were all located in a shielded compartment. The procedure of balancing the bridge for zero output voltage at power frequency consists of applying the voltage to the specimen and manipulating the bridge arms ($C_2$ and $R_2$) for a fixed known value of $C_1$ such that the oscilloscope shows zero vertical deflection. The procedure was to be repeated successively after the application of each voltage increment. When the balancing is precise, and if the pulse-less partial discharge appears, then, the oscilloscope trace opens out into a series of increasing and decreasing steps in consonance with the applied voltage. The amount of charge transferred due to the micro discharges to the measuring circuit can be worked out by applying the calibration constant. The micro discharges of small magnitudes (0.3–0.8 pC) in a series of steps could be observed on the scope in the case of plain polyethylene samples with extremely divergent electric field.

Figure 2.23 gives oscillograms of discharges (non-impulsive) in LDPE. In the case of CIGRE Model I cell, it was possible to establish ample evidence for the existence of swarming pulsive micro discharges of high repetition rate and small magnitudes as described by Karner et al. [11].

Based on the series of measurement, a very important observation could be made. The pulse-less discharges appear only when a gas phase is present in the divergent field. If such a phase is not present, then it was observed that the specimen gives rise to pulse type partial discharge and with the application of higher voltages, it breaks down with occasionally showing pulse-less partial discharges. This is understandable because, the non-impulsive discharges, by their basic physics are probably formed by collecting charges on the walls of a cavity which later annihilate by recombination of some kind giving a sort of a glow. Extending this a little further, in a system which apparently is very compact without any gaseous cavities or defects one may not see micro discharges.

After certain duration of ageing under electric stress, if any other type of defect gives rise to, for example treeing, then in the channels of the tree, it becomes possible to observe micro discharges. To support this finding further, a number of identical specimens of epoxy resin (about 50) with no artificial voids were fabricated and tested for the absence of partial discharges up to 30 kV (partial discharged $\leq$ 0.5pC).

A major point of interest to utilities is the possibility of estimating the remaining life of insulation. Diagnostics for carrying out preventive

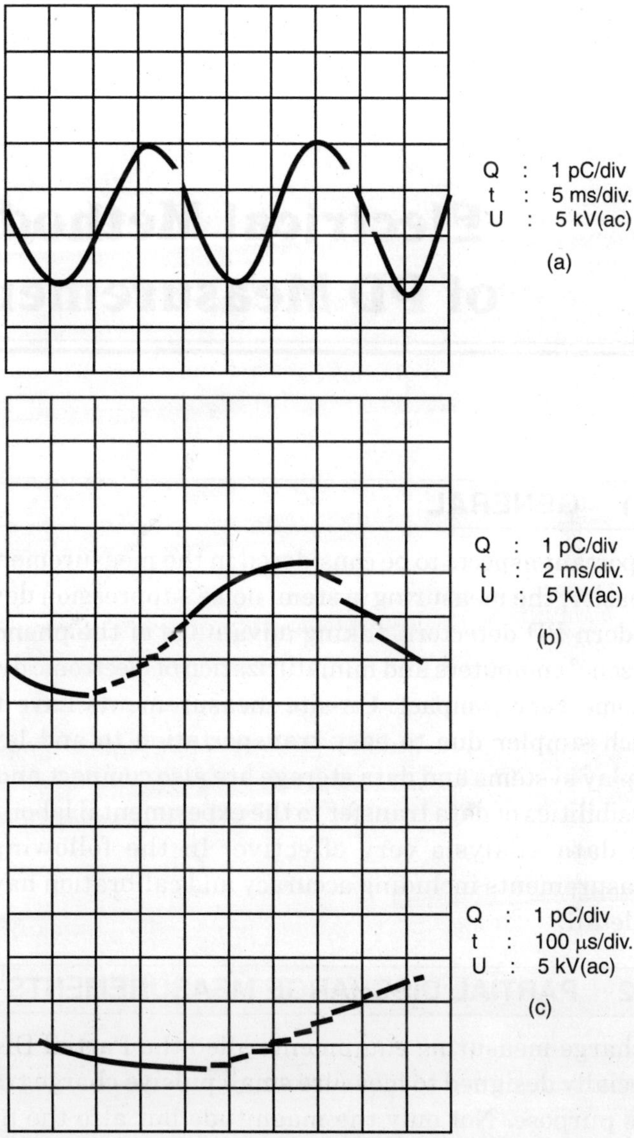


**Fig. 2.23.** Oscillograms of μ-discharges in LDPE

maintenance of apparatus insulation based on non-destructive measurements. With this in view, the research programs are oriented towards establishing a tangible connection between the time to failure and charges in the micro discharges pattern parameters (magnitude, phase and number density).

CHAPTER 3

# Electrical Methods of PD Measurement

## 3.1 GENERAL

Important aspects to be considered in the measurements are, the source capacity, the measuring system, noise suppression devices and cabling. Modern PD detectors, taking advantage of the phenomenal growth of personal computers and miniaturization of electronic devices increasingly become very compact. On-site measurements have therefore become much simpler due to easy transportation to and layout at site. The display systems and data storage are also compact and by telemetry the possibilities of data transfer to the experimental laboratories have made the data analysis very effective. In the following, all aspects of measurements including accuracy and calibration have been described in detail.

## 3.2 PARTIAL DISCHARGE MEASUREMENTS

A charge measuring equipment called the Partial Discharge Detector, specially designed to measure small pulsive charge transfer, is used for this purpose. Not only the magnitude but also the number density of charges of a known magnitude can be detected and quantified using this equipment. In an actual measurement, the charge pulse is transformed into a voltage signal, $U_Z$. This can be accomplished with the help of a specially designed impedance network, $Z$, as shown in Fig. 3.1.

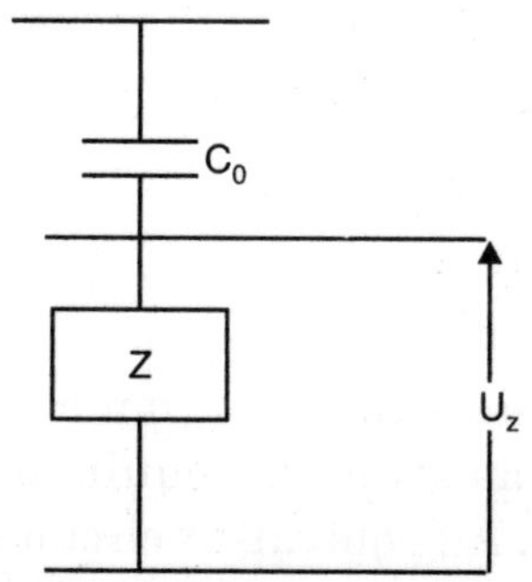


**Fig. 3.1.** Output measuring impedance for a pulsive current

The voltage, $U_Z$, across the measuring impedance, $Z$, due to a pulsive charge transfer is dependent upon the components in $Z$. If $C_0$ is the capacitance of the cavity, the ratio of the time integral over the current pulse and the cavity capacitance is the representative of the true charge due to the partial breakdown of the cavity. Formally,

$$U_z = \left(\frac{1}{C_0}\right)\int_0^t i \cdot dt \tag{3.1}$$

The charge, $q$, at the terminals of the test specimen is only proportional to the true charge but not the true charge itself. This is due to the fact that the charge transfer is considerably attenuated and the pulse distorted as it appears across the terminals. The charge, $q$, is called the apparent charge given by the Eqn. (3.2).

$$U_z = \frac{q}{C_0} \tag{3.2}$$

## 3.3 PD MEASURING CIRCUITS

PD in insulation system causes patterns of current impulses in the terminals of the electrical equipment. Detection of these impulses requires the following items:

i. PD free high voltage source for application of voltage.
ii. Measuring impedance across which voltage impulses are caused by the discharge impulses in the sample.
iii. Coupling capacitance which facilitates the passage of the high frequency current impulses and must be of at least the same order of capacitance as the sample.
iv. Amplifier

v. Detection and displaying unit
vi. Filters for noise suppression

Two basic methods of measurement of PD are:

i. Straight detection
ii. Balanced detection

Straight detection method is the most commonly employed method for laboratory testing of most of the equipment like power cables, transformers, and so on. An equivalent circuit for straight detection is shown in Fig. 3.2. Balanced detection is generally employed for minimizing the power supply disturbances and for high capacitance samples like capacitors to achieve higher sensitivity. A typical arrangement for balanced method is shown in Fig. 3.3. The essential feature of this method is the common mode rejection paradigm by which the noise in one arm of the detector is eliminated by the other.

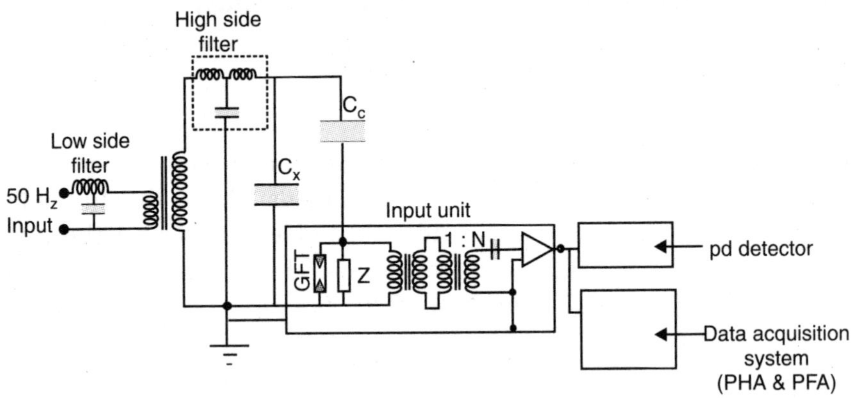


**Fig. 3.2.** Schematic showing straight detection circuit

Any PD measuring equipment seeks to measure the peak amplitude of voltage dip, $\Delta V$. If the nominal capacitance of the specimen is known, the charge associated with the pulse can be calculated. Upon a breakdown of the discharging cavity, nearly all the charge residing on the walls of the cavity in annihilated in a very short time, (< 1 μs) and this process constitutes a measurable current through the external circuit. Referring to the Figs. 3.2, 3.3 and 3.4, this current is sensed as a voltage pulse across an impedance of known characteristics, amplified and displayed on an analogue or digital device.

The PD detection based on the above circuit (Figs. 3.3 and 3.4) is called 'Common Mode' rejection principle where noise coming from two sources having common frequency and other characteristics are compared and rejected by appropriate balancing of the bridge using the lower arms of the circuit.

The essential components of PD circuit shown in Fig. 3.2 include, (*a*) High voltage source, discharge-free up to its maximum rated voltage, (*b*) High frequency (HF) filter, (*c*) High voltage discharge-free coupling capacitor, (*d*) Measuring impedance, HF matching transformer and tuned amplifier and (*e*) Display device.

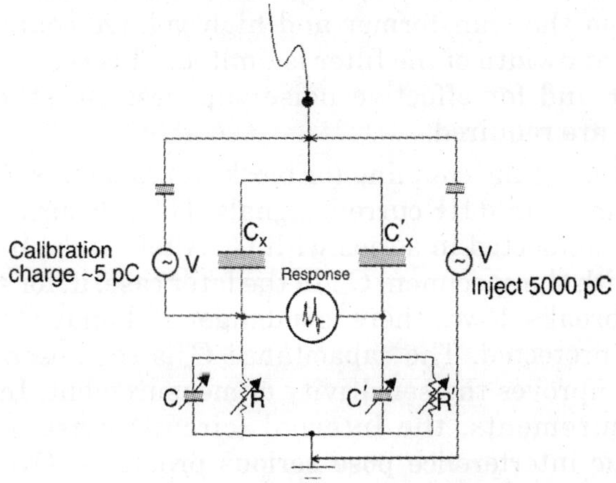


**Fig. 3.3.** Circuit showing the principle of balanced detection

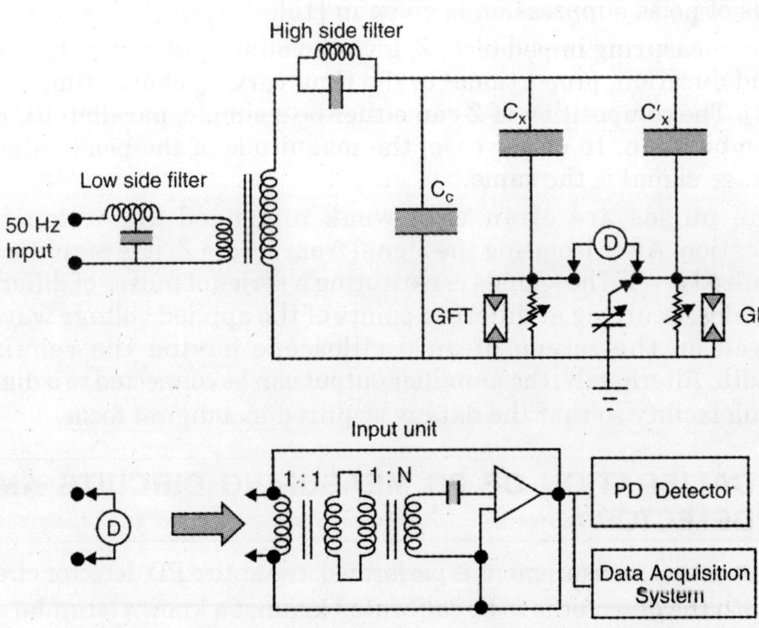


**Fig. 3.4.** Schematic showing of balanced detection circuit

The alternating voltage source is a carefully shielded testing transformer with liberally designed insulation and corona-free connecting terminals. It often happens that, current demands by large capacitive loads such as long cables, high voltage capacitors of various types and the like place serious restrictions on the testing transformers necessitating reactive compensating devices.

The HF filter is essentially an air core inductor coil, acting as a narrow band noise-suppressor, is meant to block the disturbances emanating from the transformer and high voltage connecting leads. The effective bandwidth of the filter is limited. This is only an initial or a coarse filter and for effective noise suppression, other elaborate arrangements are required.

The function of the coupling (or blocking) capacitor, $C_c$ provides a return path to measured HF current signals. The measuring impedance, $Z$, is normally connected in series with $C_c$. Alternately, it may also be put in series with the specimen, $C_x$. In the later case, if, for some reason, the specimen breaks down, there is a danger of damage to $Z$, unless it is adequately protected. The capacitance $C_c$ is so chosen as to make $C_c > C_x$. This improves the sensitivity of measurement. In performing reliable measurements, the internal (circuit) noise and external electromagnetic interference pose serious problems. Over the years, numerous improvements have been made in resolving these and associated problems to a great extent. A detailed description on the methods of noise suppression is given in [15].

The measuring impedance, $Z$, gives a voltage pulse of a short rise time and duration, proportional to the time varying charge (impulsive current). The composition of $Z$ can either be a simple, parallel- RC or a LRC-combination. In either case, the magnitude of the peak value of the voltage signal is the same.

The pulses are often very weak and need distortion-free amplification. After isolating the signal from across $Z$, it is amplified to the required level. The signals constituting a series of pulses of differing magnitudes occurring at different points of the applied voltage wave is displayed on the screen of an oscilloscope having the required bandwidth. Alternately, the amplifier output can be connected to a digital recording facility so that the data is acquired in a digital form.

## 3.4 CALIBRATION OF PD MEASURING CIRCUITS AND DETECTOR

Before an actual measurement is performed, the entire PD detector circuit along with the detector is to be calibrated against a known (standard/or equivalent) pulse source. In this process it is also possible to estimate the amount of signal attenuation, distortion and a possible mismatch in the pulse shape. Calibration circuits for charge detection are simple devices as shown in the circuits below. In Figs. 3.5 and 3.6, one of the methods of source calibration is demonstrated. When measurements are performed on-site it is often necessary to carry a portable calibrator, Fig. 3.7.

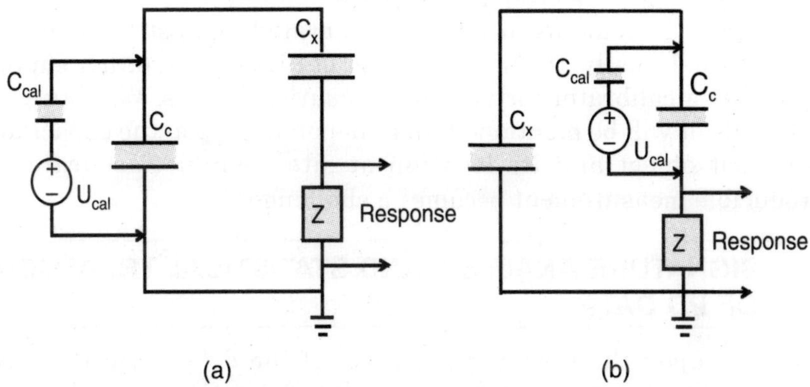


**Fig. 3.5.** Circuits for calibration of PD from a test object
**(a)** Across the coupling capacitor and **(b)** Across the test object

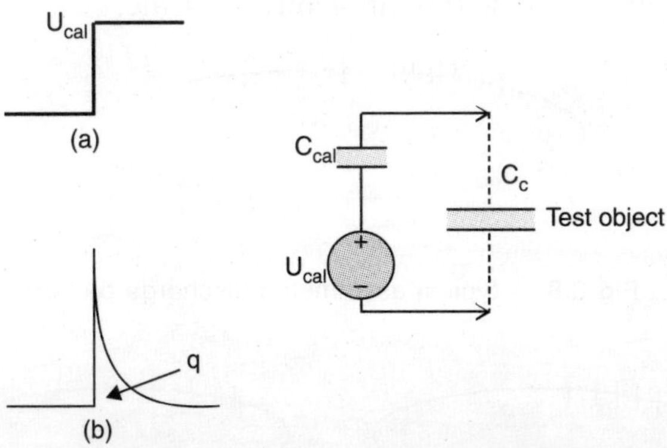


**Fig. 3.6.** **(a)** Voltage impulse produced by a square wave and **(b)** Charge displacement caused by this impulse

**Fig. 3.7.** A portable calibrator

It is to be noted that the magnitude of the voltage signal and therefore the charge measured at the amplifier terminal (refer to Figs. 3.2 and 3.4) is much less than the actual value due to circuit attenuation. Furthermore, the apparent charge, $q = C_x{*}\Delta V$ is measurable only with reference to a known charge pulse, injected at various points of the measuring circuit and measuring the attenuation.

The PD detector itself is to be calibrated against the reference charge pulse. Usually, the manufacturer of PD detector either supplies or suggests a calibrator for various measuring ranges. Without going into details, it will be mentioned, that, depending upon the capacitance of the test object and its location at site, reliable, accurate and reproducible measurement becomes a challenge.

## 3.5 SIGNATURE ANALYSIS AND STATISTICAL TREATMENT OF PD DATA

Depending upon the type and geometry of the defect, the PD leaves indelible imprints or signature or pugmarks with a high degree of reproducibility. The imprints open up vistas in enumerating the source, the ambient, and the defect types. An indication of possible imprints are shown in Figs. 3.8–3.10 to an arbitrary scale.

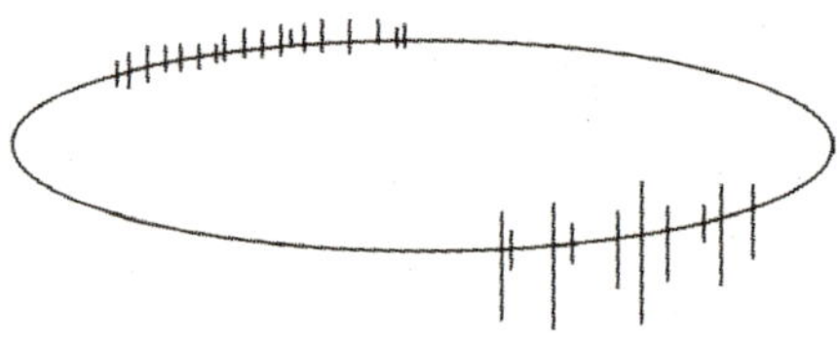

**Fig 3.8.** A typical asymmetric discharge pattern

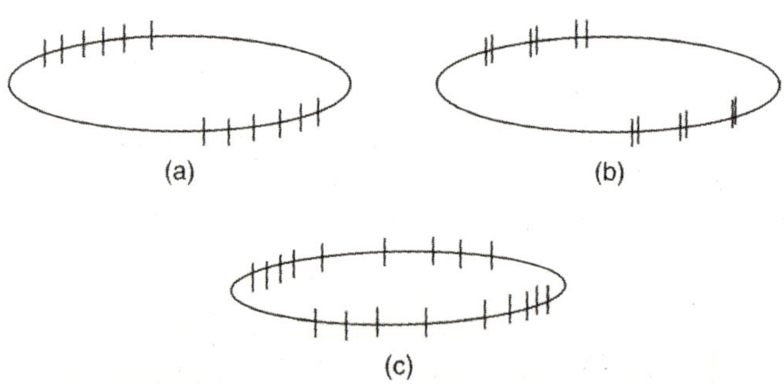

**Fig. 3.9.** Oscillograms of discharges caused by **(a)** Floating points; **(b)** Synthesized and **(c)** Actual

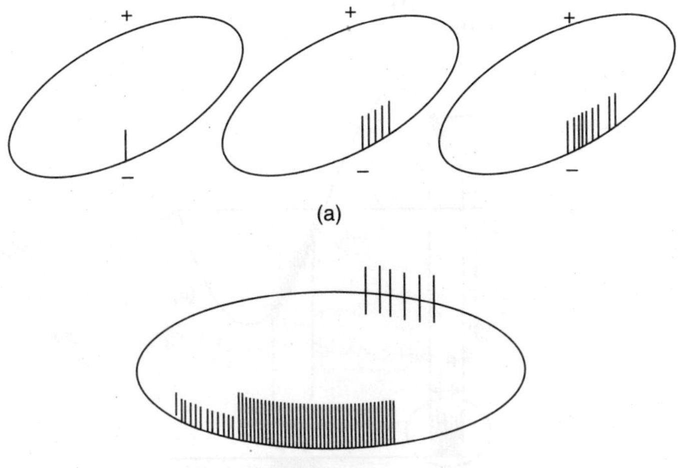

**Fig. 3.10.** Corona discharges at, and slightly above, the inception voltage. (Corona occurs at the high voltage electrode and at the negative peak). **(a)** In air and **(b)** In oil

The intention of statistical treatment to PD data is to try and extract observable quantities in a manner as to make a meaningful assessment of the degradation process. It has been observed [16] that, in addition to the number density distribution of the PD pulses, often called H (q)-n distribution, certain other statistical features are required to process the data. One such aspect is the distribution of pulses with respect to the phase angle or time window referred to a 50Hz voltage wave. This has been illustrated in Fig. 3.11.

The Authors[17] describe the following method of phase resolved PD analysis as opposed to the pulse magnitude analysis described earlier.

Step 1.

When the voltage is increased to the inception level or slightly greater, discharges start occurring as shown in Fig. 3.11(a) in a polymeric insulation. As can be seen, similar discharge process take place on both halves of the voltage wave.

Step 2.

Compensate for the 50 Hz voltage wave so that the discharge are transferred to a linear display shown in Fig. 3.11(b).

Step 3.

Quantize with convenient phase windows, preferably equal of $\Delta\theta$ shown in Fig. 3.11(c). It is suggested that the angular expanses $\theta_1$ and $\theta_2$ be quantized in such a way as to obtain not less than 10 slots, refer to Fig. 3.11(d). This means

$$\frac{\theta_1}{\Delta\theta}, \frac{\theta_2}{\Delta\theta} \geq 10$$

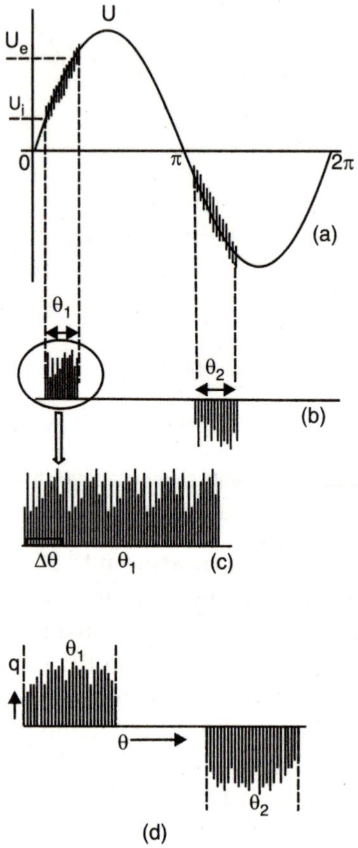


**Fig. 3.11**. A display of PD pulses on the applied voltage wave (U) **(a)** Depiction of inception and extinction, **(b)** Phase angle display, **(c)** Phase windowing in b and **(d)** Expanded version of b.

Step 4.

Compute the mean change per phase window as:

$$\left\langle q_{ph} \right\rangle = \frac{1}{n(\Delta\theta)} \sum_{j=1}^{n} q_j \; n_j \tag{3.3}$$

where $q_i$ are the charges and $n_j$ the number density $q$ charges in that window and $n$ $(\Delta\theta)$ are the number of discharge pulses in $\Delta\theta$.

Suppose a number, $N$ of the voltage cycles are considered in the analysis corresponding to each phase window. Defining the conditional occurrence (or probability) that the number of cycles, $n$ out of $N$, in which the charge magnitude $q_k$ of the $k^{\text{th}}$ phase-window equal to or greater than the mean $q$ of that window, a conditional probability statement can be written as:

$$p\left[q_k = \langle q \rangle | \Delta\theta, n\right] \tag{3.4}$$

The probability expressed in Eqn. (3.4) is called the probability density (*pdf*). A plot of pdf of the random variable $q_k$ gives the instantaneous probability of the charge per cycle. This phase resolved PD pulse frequency analysis is, in general, not symmetric. The degree of asymmetry and its sign will, in some way, convey whether the insulation is degrading rapidly or otherwise. In polymeric insulation, the phase distribution of pulses is often peaked to the right (positive skewness) when the insulation is new. As it ages, slowly, the distribution becomes nearly symmetric (zero skewness) later, in advanced ageing situation skewness becomes negative. It is possible to arrive at a reason assessment of the state of the insulation by following the skewness of the phase distribution of PD pulses. It is to be stated that, the $q$–$\phi$ is seen, by many authors, to be more sensitive to the amount of ageing than the $q$–$n$ distribution skewness.

For the sake of brevity only salient features of these statistics are considered. The kurtosis of a distribution gives an indication of the density of probability around the nodal value. A sharp peak (large peakedness or kurtosis) around the nodal value indicates a concentration of the process parameter (in the present case PD). Kurtosis is considered to indicate only a qualitative picture of the random process. Other than this, its further consideration in insulation ageing studies is not warranted. In insulation ageing studies, the observable parameters '$k$' and $S_k$ ($k$-statistics as they are generally called) give a very useful information on the status of insulation.

Formally, the skewness and kurtosis are defined as:

$$S_k = \left(\frac{1}{n}\right) \sum_i f_i (x_i - \bar{x})^3 \tag{3.5}$$

$$k = \left(\frac{1}{n}\right) \sum_i f_i (x_i - \bar{x})^4 \tag{3.6}$$

The random variable (RV) "**X**" can be any physical quantity, in the present case, the PD magnitude. The probability density $f(x)$ of random variable **X** is the probability that the $i^{th}$ discharge magnitude and $X = x_i$, a measured value.

Referring to Fig. 3.12, for (1) and (2), the random variable $X$,(PD) has a large number of $x_i$'s greater than the mean and hence $S_k > 0$. For (4) and (5), $S_k < 0$.

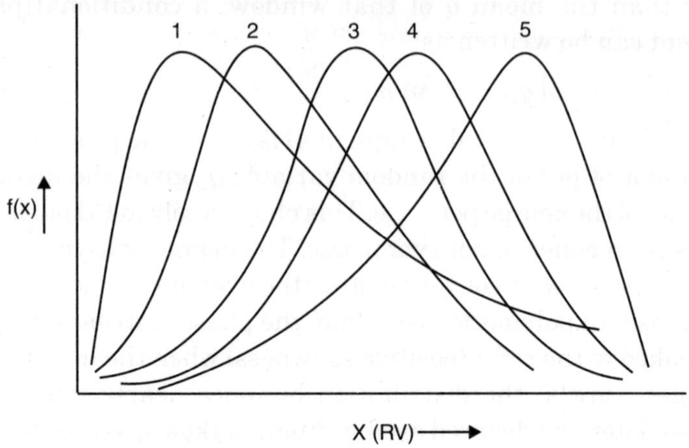


**Fig. 3.12.** Demonstration of skewness of probability distribution. (1) & (2) Skewed to the "Right" ($S_k > 0$), (4) & (5) Skewed to the "Left" ($S_k < 0$) (3) Symmetric distribution ($S_k = 0$)

The meaning of this happening is that in the beginning of the ageing process the number density of small discharges ($q \leq 10$ pC) is larger than those of the large discharges. As the insulation ages slowly, the number of larger discharges ($q \geq 100$ pC) increases. In other words, either the number of discharge sites is increasing or the energy dissipated at the site of already existing discharges is increasing indicating that the material is undergoing rapid destruction. Between these extremes, there is a regime, "the normal regime" at which the insulation is stabilized against the applied stress and the surroundings so that the insulation has successfully passed the infantile mortality (early failures due to gross defects) stage and is in a normal status of working.

## 3.6 INDIRECT ELECTRICAL MEASUREMENT

In the electrical method of discharge detection, the required signal is often heavily infested with noise of various origins some of which happen to overlap the signal frequency spectrum. This makes signal retrieval very difficult and also renders signal to noise ratio, (s/n) very poor. Further, depending upon the capacitance of the test object, the test voltage source becomes very bulky thereby making on site measurements quite expensive and laborious. For this and several other reasons, non-electrical and indirect methods of PD measurement have come into vogue. These testing techniques do not measure the apparent charge directly, but, use certain physical side effects produced by the discharges to detect and quantify them. Obviously there is a need to establish a correlation and functional relationship between the measured physical effects and PD quantities. In practice, however, non-electrical techniques are used more for detection and possible location of discharges than for quantification. Among such methods, UHF and acoustic detection merit consideration here.

A burst of partial discharges in a liquid or a rarefied medium is accompanied by electromagnetic, (EM) radiation in different modes, with frequencies in the (UHF) range, of 2...5 GHz. The ensuing UHF radiation can be coupled to a carefully designed external (capacitive) coupling device. The voltage output thereof can be recorded and displayed on a spectrum analyzer/recorder. This technique is particularly useful in the diagnosis of particulate defects in Gas Insulated Substations (GIS) and large rotating machine stator insulation.

## 3.7 UHF SENSORS AND HF CT

Among the non-conventional method of PD measurements having electrical overtones are High Frequency (HF), Very High Frequency (VHF) and Ultra High Frequency (UHF) techniques [18]. The characteristic frequency ranges are indicated in Table 3.1.

**Table 3.1.** Frequency ranges used in different measuring techniques

| Electro magnetic technique | Frequency ranges, MHz |
|---|---|
| HF | 3 – 30 |
| VHF | 30 – 300 |
| UHF | 300 – 3000 |

The HF and VHF frequencies are indicated in power cables where the electromagnetic waves are sensed by inductive or capacitive sensors. The UHF frequency range in particular is adopted in gas insulated systems. Table 3.2 indicates the high frequency methods applicable to power equipment.

**Table 3.2.** Frequency ranges used in different power equipment

| Electro magnetic technique | Cables | Transformers | GIS | Generators |
|---|---|---|---|---|
| HF | + | – | – | + |
| VHF | +/– | + | + | + |
| UHF | +/– | + | + | – |

As applied to GIS, with a working frequency range of PD between 0.3 and 3 GHz, capacitive sensors of antenna type are normally employed to capture the PD signals. Two types of capacitive sensors, shown in Figs. 3.13(a) and (b) are used in GIS. The essential features of capacitive sensors are that they are located on the body of the GIS across a specially designed flange. The central shielded electrode is normally in the form of a disc, mounted over a loosely coupled hermitically sealed collar and gas tight plug arrangement through which the connecting lead is taken out. The disc interacts with the radial component of the electromagnetic wave produced by a PD pulse in the gaseous medium. The changes in the electromagnetic wave input as sensed by the capacitive sensor are lead into appropriate measuring circuits for measurement and display.

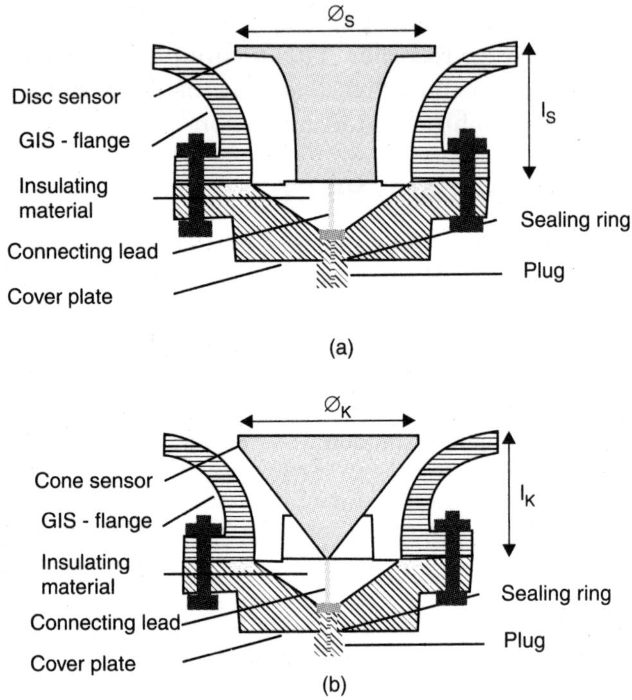


**Fig. 3.13**. Pictorial view of UHF PD capacitive sensors [18] **(a)** Cylindrical collar sensor and **(b)** Conical collar

High frequency current transformer (HFCT) sensors of the Rogoskowki coil type [18, 19] have long since been used for fast transient measurements. Among the more important applications of this method of sensing are, measurement of atmospheric lightning currents, very fast transient over voltages (VFTO) occurring in gas insulated systems due to electron burst and in UHF PD in GIS. The principle of Rogoskowki coils is essentially based on Maxwell magnetic coupling of fast time varying electromagnetic signals propagating along a straight conductor, Fig. 3.14. Referring to Fig. 3.14, a circular ferrite toroid over which a copper coil is wound represents the secondary of a current transformer. The primary consists of a straight conductor, for example, the current carrying part of

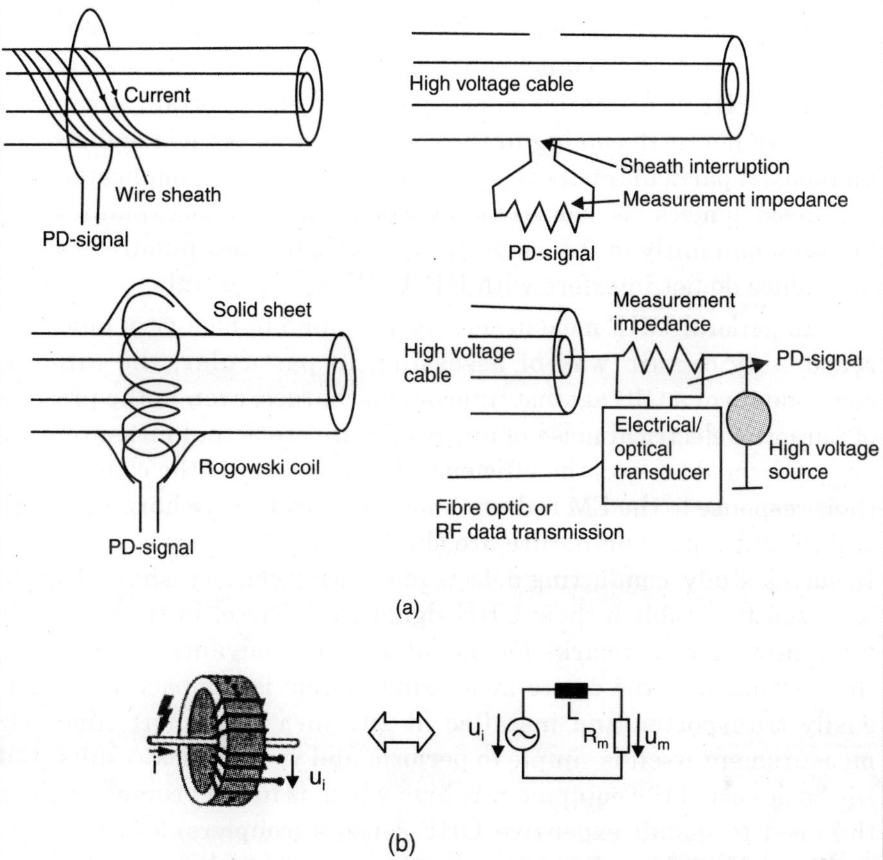


**Fig. 3.14.** HFCT for PD measurement (Frequency range: 1- 40 MHz) [18] **(a)** Rogowski coil and associated coils and **(b)** Rogowski coil and its equivalent circuit

a cable. PD occurring in the insulation induces transient voltages/ currents in the central conductor which travels along as a transverse magnetic wave (TM wave). The interaction of this wave component to the secondary coil of the magnetically coupled ferrite core (the CT) produce time varying voltage pulses between the terminals of the coil. These voltage signals are in some way proportional to the PD and are as such taken as a measure of the PD activity. As can be seen from the principle of measurement there are several epochs of attenuation of the signal as well as frequency and wave form distortion. The method becomes advantages and simple only when the PD signals have large magnitudes.

In an actual measurement a split core CT becomes a necessity.

Care should be taken to minimize the air gap in the ferrite core so that the reluctance of the magnetic circuit is not disturbed. The number of turns in the winding is important as it determines the signal strength.

The UHF and associated measurements enjoy certain advantages which are not forth coming in the conventional electrical and acoustic methods. A particular importance is the irrelevance of complicated noise suppression methods. Noise of known origin as shall be explained later are predominantly in the range of a few kHz to a few hundreds of kHz and hence do not interfere with HF–UHF measurements.

In general UHF measurements are found to be a very quick and reasonably certain way of assessing, in particular, the different components of a GIS against internal discharges, even in the presence of parasitic electrical noise of any origin. Before actual measurements are performed on site, the efficiency and precision of the couplers and their response to the EM radiation produced by the discharging defects of different types and nature are checked on laboratory scale models. In such a study, conducting defects of known geometry, shape and size are used to establish their UHF signatures. These, in turn, serve as references or benchmarks for use at site. The advantages offered by this method are, the entire measuring system is compact and can be easily transported and installed at site in a very short time. The measurement itself is simple to perform and yields quick results. But, the basic cost of the equipment is high which is further compounded by the need to install expensive UHF sensors (couplers) spread over a number of locations. Furthermore, the analysis and interpretation of the data needs experience and expertise.

A part of the energy due to partial discharges occurring in power apparatus is dissipated in the emission of signals in the audible frequency or (acoustic) range, besides UHF signals. Acoustic waves of nearly spherical wave front at the point of their inception, become rarified as they traverse the medium in which they are generated. These waves interact with the walls of the metal enclosure and set up standing and traveling waves in the matrix in a highly complex way. Avoiding details, it can be shown that depending upon the structure and the elastic properties of the material from which the enclosure is made of, the wall vibrates, predominantly in a direction normal to it. The elemental displacement, acceleration, velocity and other elastic parameters of the enclosure can be measured using appropriate sensors as electrical output (voltage signals). Since the primary cause of the vibrations is the partial discharge energy and hence the apparent charge, $q$, the measured elastic parameters are functions of $q$. Many of the remarks made in the preceding paragraph are relevant here too in that only a qualitative assessment of the PD is possible. For determining the magnitude of apparent charge, other and more direct methods are to be resorted to.

## 3.8 MEASUREMENT OF DISCHARGES UNDER DIRECT VOLTAGES

Under direct voltages the partial breakdown mechanism due to defects is, physically, very different than under alternating voltages. The difference are due, in part to the non-zero crossing event of the voltage waveform under DC. The Paschen curve appears formally similar under AC and DC, however, the Paschen minimum is elevated considerably from the corresponding AC value. It is well known that DC PD is understood in terms of the current pulses and hence expressing this as an equivalent charge is sometimes seen to be incorrect. In view of this an equivalent circuit involving resistivities in addition to capacitances are considered as described below.

### 3.8.1 Electrical Methods

The detection of the electrical pulses produced by partial discharges tends to be more convenient and sensitive than the non-electrical methods. Typical discharge detection circuits for both alternating and direct voltage conditions are shown in Fig. 3.15. $X_D$ represents the detection impedance, $C_a$ is the test specimen, and $C_k$ is the capacitance in parallel with the test object and the detection impedance. $C_k$ provides a low impedance path for the high frequency current of a discharge. The detection impedance can be connected to the low-voltage terminal of $C_k$ rather than $C_a$. The response of the detection impedance to a discharge in the test specimen is amplified, measured, displayed, and counted.

Under alternating-voltage conditions, discharges can occur regularly during every cycle, and it is possible to measure the energy dissipated by the discharges per cycle [20]. However, for direct voltages, the discharges are much less frequent and occur randomly with time. Consequently, direct voltage discharge-detection systems do not measure the discharge energy but detect and measure the individual discharges or discharge current.

#### *3.8.1.1 Detection Impedance*

The detection impedance can be either a resistive-capacitive (RC) or resistive-capacitance-inductive (RCL) type, Fig. 3.15(a). The RC type gives

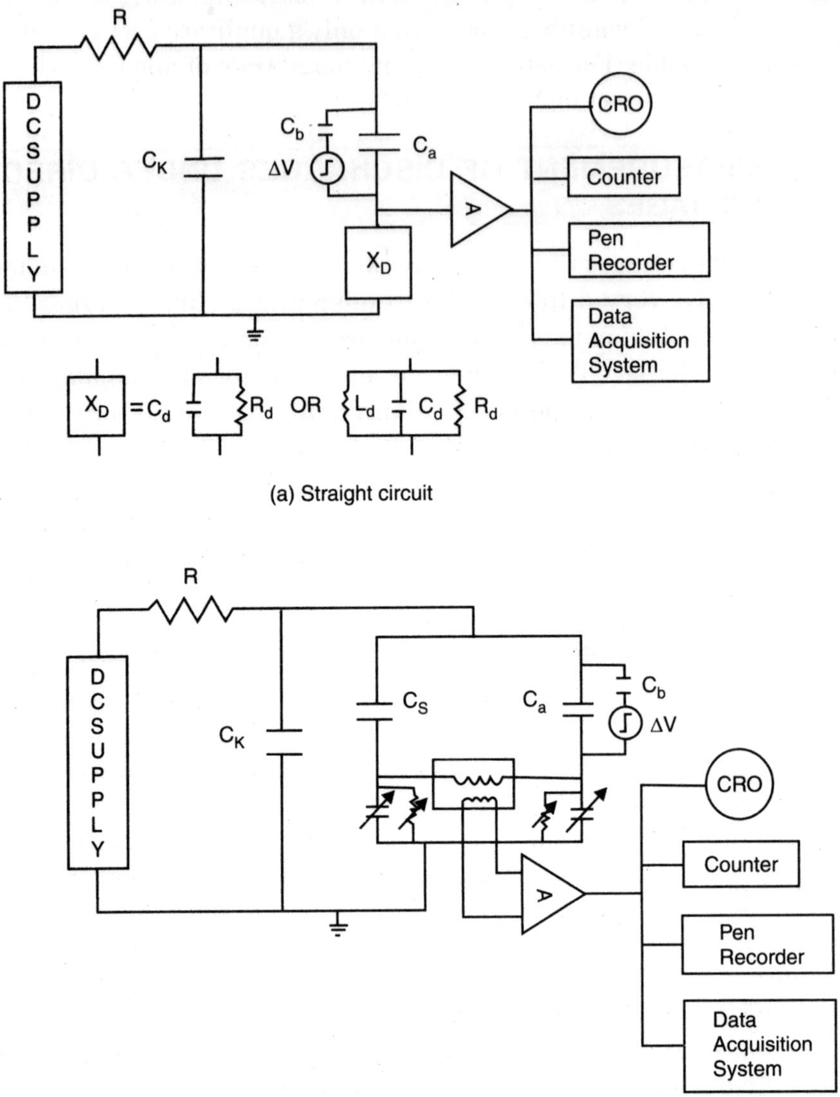


(b) Balanced circuit

**Fig. 3.15.** Typical AC/DC discharge detection circuits

a unidirectional-pulse output, whereas the output for the RCL type is a damped oscillation. The voltage across a RC- type detection impedance for a discharge of magnitude, $q$, in the test specimen is given by:

$$S = \left[ \frac{q}{C_a + C_d + \left\{ \frac{C_a\, C_d}{C_k} \right\}} \right] \times \exp\left[ \frac{-t}{R_d\left( C_d + \left\{ \frac{C_a C_k}{C_a + C_k} \right\} \right)} \right] \tag{3.7}$$

The voltage across an RLC-type detection impedance is

$$V_a = \left[ \frac{q}{C_a + C_d + \left\{ \frac{C_a\, C_d}{C_k} \right\}} \right] \times \exp\left[ \frac{-t}{2R_d\left( C_a + \left\{ \frac{C_a C_k}{C_a + C_k} \right\} \right)} \right] \cos \omega t \tag{3.8}$$

where;

$$\omega = \left( \frac{1}{L_d p} - \frac{1}{4R_d^2 p^2} \right)^{1/2} \tag{3.9}$$

and

$$p = C_d + \frac{C_a C_k}{C_a + C_k} \tag{3.10}$$

The sensitivity can be increased by decreasing $C_a$ and $C_d$ and making $C_k$ as large as possible.

#### *3.8.1.2 Discharge Measurement*

The output pulses from the detection impedance are amplified and the amplified signals processed in different ways. A wide-band amplifier is used with a RC-detection impedance and a narrow-band amplifier is coupled to a RLC–detection impedance. The centre frequency of the amplifier is matched to the resonant frequency of the detection impedance. The amplified signals are suitably shaped so that they can be counted.

### 3.8.2 Calibration

Discharge-detection circuits are usually calibrated so that the output can be measured directly in pico Coulomb (pC). To calibrate the detector, a known pulse of charge from a calibrator is injected into the test circuit and the response of the detector measured. By comparing the detector responses due to a calibration pulse and a discharge within the test object, and considering the detection-circuit parameters, the magnitude of the discharge within the test object can be evaluated. The calibration charge can be injected across the test object, with no high voltage applied, or across the detection impedance so that calibration pulses can be displayed with high voltage applied to the test object.

### 3.8.2.1 Across the Test Object

The circuit is shown in Fig. 3.16(a). The calibrator that supplies voltage pulses, $V_q$, through a capacitor, $C_q$, is connected across the test object, $C_a$. The magnitude of the calibration charge can be varied by changing $V_q$ or $C_q$. The calibration capacitance, $C_q$, is usually much smaller than $C_a$ so that the change in potential at the terminals of $C_a$ due to the calibration pulse is $C_q\ V_q/C_a$, and the voltage across the detection impedance is given by, $q = V_q\ C_q$.

### 3.8.2.2 Across Detection Impedance

The calibrator is connected across the detection impedance as shown in Fig. 3.16(b). The voltage across the detection impedance due to a

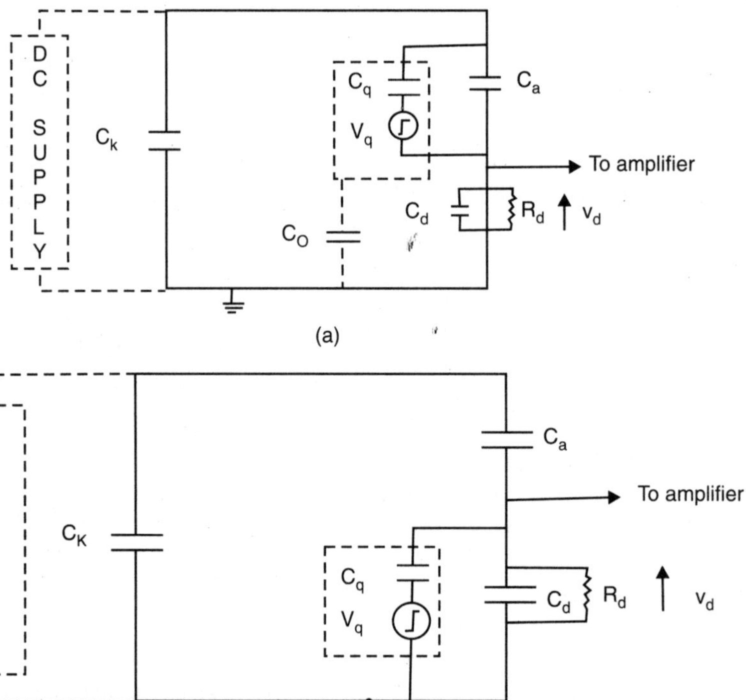


**Fig. 3.16.** Calibration of straight discharge detection circuit. **(a)** Calibrator connected across test object and **(b)** Calibrator connected across detection impedance

discharge in the test object is given by the Eqn. (3.7). If the voltage pulses are equal in magnitude, then magnitude of apparent charge, $q$, is given by

$$q = Q\left(1 + \frac{C_a}{C_k}\right) \tag{3.11}$$

## 3.9 MEASURING CIRCUITS FOR MICRO DISCHARGES

The measuring circuits used in the investigations reported here was developed by KEMA, based on a similar arrangement by Lemke[10]. Actually, the original circuits were devised by Konig and Kubler[12, 21]. In principle, the measuring circuit is an integrating bridge of the type described by Dakin[22]. Figure 3.17(a) shows the essential circuit and Fig. 3.17(b) is the complete set-up.

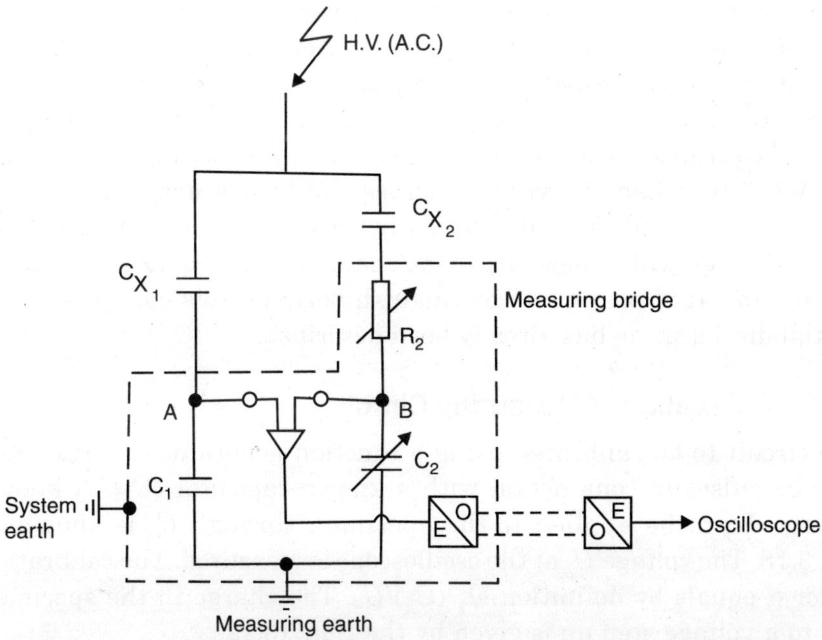


**Fig. 3.17(a).** Principle of measuring bridge

$C_1$ is a low capacitor into which the charges created by the glow discharges occurring in $C_{X1}$, are drained. These charges are added in step (integrated) over a full voltage cycle as the charges follow the applied voltage. The integrated charges (voltage steps) are then fed to a very high input impedance measuring device. This essentially consists of an impedance converter giving an input impedance of greater than $10^{10}$ Ù and a series of post-amplifiers of variable gain with appropriate feedback.

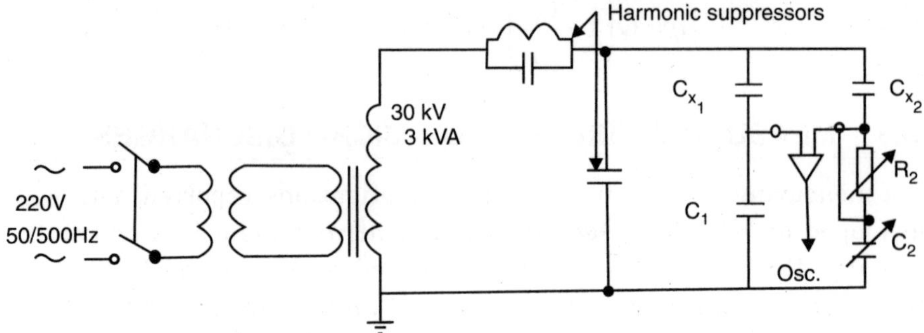


**Fig. 3.17(b).** Schematic of the measuring set up

The other branch of the circuitry consists of $C_{X2}$ and an identical capacitor to the specimen having approximately the same loss and capacitance. $C_{X2}$ is another (reference) sample or a compressed gas standard capacitor. Any difference between the loss in the specimen (*S*) $C_{X1}$ and $C_{X2}$ can be balanced out by the variable resistor $R_2$ and capacitor $C_2$. Actually, when the voltage across the bridge arms A-B due to balanced in magnitude and phase, any voltage appearing across these points later on will be measure of the charge developed in the specimen due to μ-*d*. It is possible to distinguish between this charge and the partial discharge as has already been described.

### 3.9.1 Calibration of Measuring Circuit

The circuit to be calibrated using a function generator *G* giving step voltage pulses in conjunction with a known capacitor ($C_S$). A known voltage $U_T$ to be applied to the specimen through $C_S$ is shown in Fig. 3.18. The voltage $U_o$ at the oscilloscope is measured. The calibration factor α equals by definition $Q_T\,(C_X)/U_o$. The charge in the specimen due to a voltage step up is given by the Eqn. (3.12).

$$Q_T\,(C_x) = U_T \times \left\{\frac{C_X C_S}{C_S + 2C_X}\right\}, \text{ (note } C_{X1} \approx C_{X2} = C_X \qquad (3.12)$$

Defining a quantity, α, as:

$$\alpha = \left(\frac{U_T}{U_o}\right) \times \left\{\frac{(C_X C_S)}{C_S + 2C_X}\right\} \qquad (3.13)$$

where $U_T$ and $U_o$ are known values.

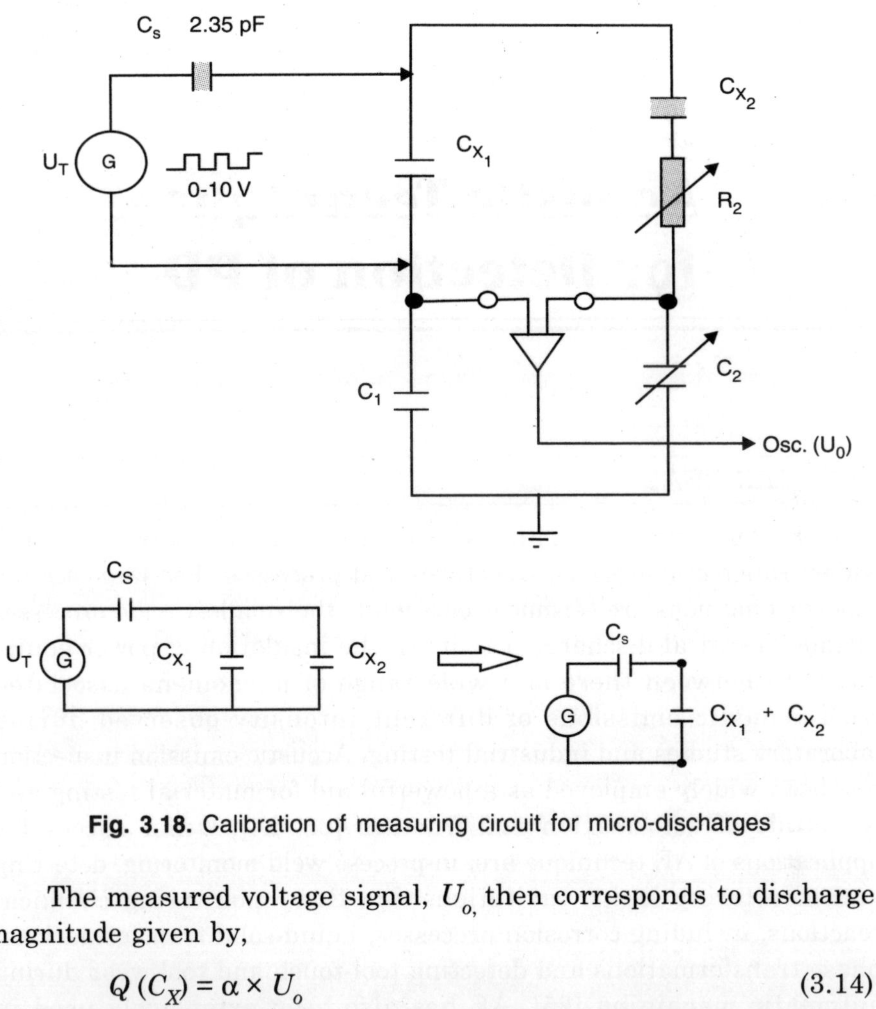


**Fig. 3.18.** Calibration of measuring circuit for micro-discharges

The measured voltage signal, $U_{o,}$ then corresponds to discharge magnitude given by,

$$Q\,(C_X) = \alpha \times U_o \tag{3.14}$$

It is essential that the test circuit should be calibrated after every change that is made in the circuit because the circuit calibration may be different for each case.

Standardizing bodies such as IEC, IEEE and CIGRE have given extensive guidelines for conducting PD measurements which include calibration, measuring circuits, approximations, accuracies and limits for PD for various equipment. Since this information is diverse and exhaustive, the more important one's are given in Appendix 2 for reference.

CHAPTER 4

# Acoustic Technique for Detection of PD

## 4.1 GENERAL

Acoustic emission (AE) is a natural phenomenon occurring in the widest range of materials, structures and processes. The largest scale acoustic emissions are seismic events while the smallest-scale processes include electrical discharge activity in the insulation of power equipment. In between there is a wide range of phenomena associated with acoustic emissions of different intensity observed during laboratory studies and industrial testing. Acoustic emission inspection has been widely employed as a powerful aid for material testing and the study of deformation and fracture [23, 24]. Other successful applications of AE technique are, in-process weld monitoring, detecting and monitoring leaks, cavitations and flow, monitoring chemical reactions, including corrosion processes, liquid-solid transformations, phase transformations and detecting tool-touch and tool-wear during automatic machining [25]. AE has also been extensively used in aerodynamics to study fatigue cracks, corrosion, composite structures etc., in aircraft and aerospace [26]. Acoustical detection also has been tried in the early days of electrical engineering design of high voltage equipment, with the help of microphones and ultrasonic transducers [27]. Subsequently, acoustic testing has been accepted as one of the well-proven and widely accepted Non-destructive testing (NDT) methods for electrical power equipment. Detection and location of partial discharges in power transformers and gas insulated systems by acoustic emission techniques has been employed by the equipment manufacturers and users for a long time [28–58].

## 4.2 EVOLUTION OF ACOUSTIC TECHNIQUE

Acoustic emission or emission of sound waves occurs in nature as a consequence of time varying rarefactions and condensations in the body of a material medium through which it traverses. Every acoustic wave is associated with a finite energy and the interactions of this energy with the structure of the material (in the case of solids, the lattice structure) causes the elementary particles thereof to vibrate with respect to their equilibrium positions. The acoustic waves due to these vibrations can travel in different directions, however, the principal direction of propagation is the initial direction of the acoustic signal, the longitudinal wave. In the presence of neighbouring particles, wave scattering occurs where by secondary or transverse propagation is initiated. As examples, acoustic emissions due to seismic events, lightning and thunder and so on can be quoted.

Discharge of electricity due to trapped charges in the body of electrical insulation (inside gas filled cavities, called defects) in power equipment, also releases measurable amounts of acoustic emissions, though of considerably lower magnitudes. Detection and measurement of acoustic emission can serve as a powerful tool in material testing and studying the defects and internal deformation like cracks, fissures, fatigue related discontinuities and localized fracture in material medium. Being non-destructive and non-invasive (diagnostic), AE techniques are among the choicest methods for defect detection in metals and other structures.

Acoustic emissions from the electrical discharge source were detected as early as 1939. Since then, ultrasonic techniques have been extensively used for the diagnostic tests during both impulse and power frequency voltage tests. In the early 1950's, acoustic sensors introduced inside a transformer tank detected low frequency ( 30 – 200 Hz) acoustic emission indicating a possible core vibration due to incorrect clamping of the core assembly. In 1977, acoustic emission from electrical discharges within power transformers was measured using an ultrasonic sensing system. The acoustic emission frequency of partial discharge varies in the range of 20 kHz to 500 kHz. In the presence of solid dielectric materials in partial discharge gap, the acoustic emission frequency is about 20 kHz to 80 kHz: while the acoustic emission frequency of partial discharge in oil gap may be higher than 200 kHz [59]. An improved method of on-line detection of partial discharges was attempted in the same year by Black Burn [60].

Application of AE technique to PD measurements in power capacitors with large capacitance values (0.5 μF to ~ 20 μF) is the only possible method as the conventional electrical method cannot be employed for such high capacitance samples due to power source constraints. The Authors have applied AE technique for assessing insulation degradation during laboratory testing of power capacitors [61, 62, 63, 64]. AE technique has been employed as one of the diagnostic tools for on-line testing of oil filled equipment like power transformers and reactors of voltage rating up to 400 kV [48, 65]. Partial discharge detection in substation equipment including Gas Insulated System for localizing the defective sites has long been used in view of its versatility for such applications.

The International Electro-technical Commission, IEC (working group WG 14) has been set up by IEC TC 42 to look into the measurement of PD by electromagnetic and acoustic methods. A new IEC standard, (IEC 62478) for Acoustic and Electromagnetic PD measurements is in an advanced stage of adoption [55]. Also, IEEE trial-use guidelines have been published in "IEEE Std C57.127-2000: Trial-use guide for the detection of acoustic emissions from partial discharges in oil immersed power transformers" [70].

## 4.3 ACOUSTIC TECHNIQUE FOR PD MEASUREMENT

The conventional electrical PD detection method is the most widely employed method for almost all types of electrical equipment like, cables, transformers, gas-insulated systems (GIS) and is carried out generally in accordance with IEC 60270. Electrical PD measurement has certain limitations particularly, when applied to test objects with large capacitance and for on-site testing. The more important limitations are:

i. Requirement of huge electrical apparatus and elaborate arrangements like PD free voltage source and blocking capacitor.
ii. Elaborative and expensive electromagnetic shielding for laboratory measurements.
iii. Site measurement demands shunt down of at least a part of the power station together with the necessity of disconnection of the test object from the circuit.
iv. Lack of control over background noise.
v. Lower sensitivity for measurements on long lengths of UG cables and power capacitors.
vi. Difficulty in localizing the PD sources.

In view of these drawbacks, new test techniques such as electromagnetic and acoustic methods have come into vogue in the recent past. These non-conventional techniques are relatively new techniques and are gaining importance as they can be applied on-line for on-site PD diagnosis. With great advancement in data acquisition and computational techniques and sensing techniques, the non-conventional methods are gaining greater acceptance, particularly for localization of defective sites. More recently, acoustic technique is gaining increased popularity due to certain distinct advantages it has over other methods, as under.

i. Non-destructive and non-invasive in nature.
ii. Easier and quicker instrument set up at site.
iii. No disruption of power system operation is required.
iv. Possibility of localization of acoustic source.

## 4.4 Principle of Acoustic PD Detection

PD activity is associated with instantaneous release of energy along with other micro and macroscopic processes. A fraction of the released energy heats the material adjacent to the PD location creating a small explosion, which excites a sound wave. The wave propagates through material media enclosing the region of emission, before reaching the enclosure of the electrical equipment. High frequency AE sensors mounted on the walls of the container pick up these vibrations of the surface wave and convert it into electrical signals, which could then be captured by appropriate instrumentation and analyzed by a processing system. The shape of the detected signal depends on the source, the propagation media, the detection equipment, and the AE sensors [66]. The principle of AE technique is schematically shown in Fig. 4.1.

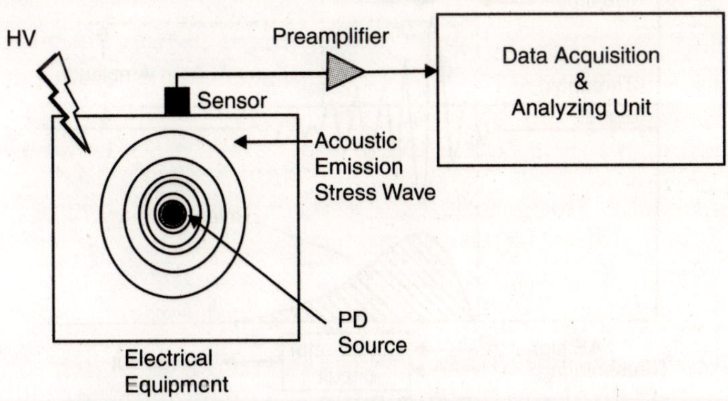


**Fig. 4.1.** The principle of AE technique (Schematic)

## 4.5 AE SIGNAL PARAMETERS

Acoustic signals are measured by several parameters like amplitude, duration, energy, rise time, counts, frequency, signal strength and so on. Standard terminologies for acoustic signal parameters as per ASTM E 1316-02 are listed in Appendix 3. For digital and analogue quantification of AE data, following are the generally accepted parameters.

i. Number density of discharges above a threshold.
ii. Signal amplitude above the set threshold.
iii. Pulse duration and rise time.
iv. Energy in terms of the area (in appropriate units) under the rectified signal envelope.

These parameters are succinctly depicted in Fig. 4.2. In many situations, the signal amplitude is, by far the most important parameter considered as a measure of the acoustic activity in question as it directly determines the detection capability of AE signal. A software-based AE system can produce many types of graphic displays between measured signal parameters. In some of the advanced AE systems, the operator's task is rather limited only to oversee the observations during the test. Post–measurement processing is effectively performed in any desired way by custom built software. Also, the acquired data can be refined, filtered and redisplayed in any manner required. The AE system can generate many real time displays of the data in the memory simultaneously even when the experiments are running.

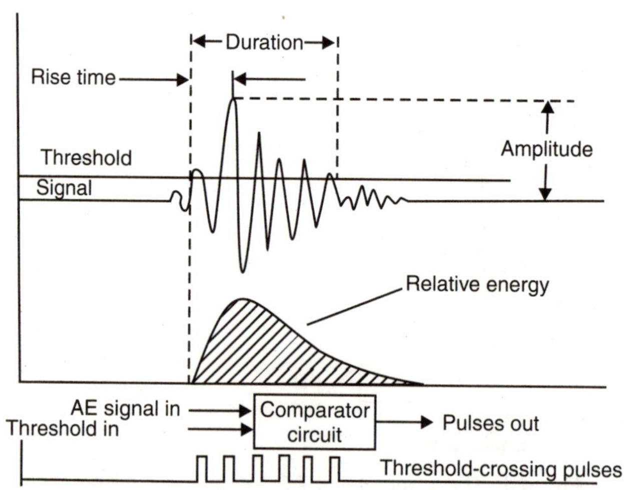


**Fig. 4.2.** Depiction of signal parameters

## 4.6 ACOUSTIC EMISSION SENSORS (TRANSDUCERS)

The acoustic signals from the electrical apparatus are converted into electrical signals by the sensors, amplified to required voltage levels by the preamplifiers. After pre-amplification, the signal is transmitted to the measurement module, where it is further amplified and filtered to improve the signal noise (S/N) ratio. The process of quantification is accomplished by a comparator circuit, which generates a digital pulse of a chosen magnitude whenever the AE signal exceeds a fixed threshold voltage. A superposition of the comparator over the AE pulse enables one to know its magnitude. The relationship between signal, threshold, and threshold-crossing pulses is shown in Fig. 4.2. The operator usually sets the threshold level, which is a key variable that determines test sensitivity. Thus detection involves the measurement of key parameters of AE signal that crosses the threshold as shown in Fig. 4.2. A digital description of AE signal is generated by the front-end hardware and is passed on to a computer system, which provides data storage, a variety of graphical displays, and replay for post-test.

The choice of appropriate sensors is a very important aspect of AE testing. The AE sensor is essentially a piezoelectric crystal which transforms the mechanical (compression) signal into an electrical (voltage) output. The piezoelectric sensor is a thin disk of piezoelectric crystal (a ceramic) metallized on both faces for intimate electrical contact and incorporated into a short metal cylinder to provide shielding from electromagnetic interferences. The ceramics commonly used in AE sensors are titanates and zirconates of Barium, molded to the desired shape and fired in a high temperature kiln. The ceramic material is then made piezoelectric by appropriate charge poling. This process involves heating the material to temperatures above its Curie point while under a strong electric field. The crystal is then housed in a metal enclosure with a wear plate and a connector, as shown in Fig. 4.3.

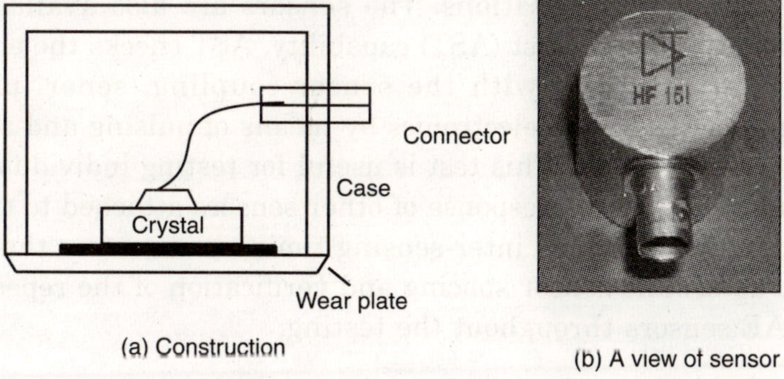


(a) Construction

(b) A view of sensor

**Fig. 4.3.** A typical AE sensor

To take advantage of the extreme sensitivity of the piezoelectric sensor, it must be attached to the material under observation in such a manner as to receive maximum possible acoustic energy with minimum loss. The physical bonding of the measuring surface of the crystal is greatly improved by using a thin film of grease like silicon at the interface. The sensors are more securely fixed to the surface of ferro-magnetic type metal frames by a magnetically actuated holder for mechanical clamping. The sensor is excited by the stress waves impinging on its face and it delivers an electrical signal to a nearby preamplifier and then to the main signal-processing equipment.

### 4.6.1 Sensor Accessories

For performing on-site testing at remote locations from the data acquisition location, suitable accessories which meet the characteristics of the AE sensors are to be selected. The important accessories are (*a*) cables with proper shielding and matching impedance, (*b*) pre-amplifiers, (*c*) line drivers, (*d*) magnetic holders where sensors are to be mounted on ferro-magnetic type metal frames, and (*e*) couplant to achieve highest contact between sensor surface and test object surface.

The signals generated by the sensors are amplified with the help of a pre-amplifier which is placed close to the sensor so as to minimize pickup of electromagnetic interference. The preamplifier has a wide dynamic range and can drive the signals over a long length of cable so that the main instrumentation can be placed hundreds of meters away from the test object, if required. Typically, the preamplifier provides a voltage gain of between 40 and 100 (15dB and 40 dB). Some of the AE sensors are available with integrated preamplifier incorporating two functions into one unit thereby decreasing the set up time for field applications. The sensors are also available with optional Auto Sensor Test (AST) capability. AST checks the entire AE signal chain starting with the sensor coupling, senor, pre-amp, cabling and AE system electronics by means of pulsing and receiving a simulated AE burst. This test is useful for testing individual sensor coupling, verifying the response of other sensors attached to the same equipment, establishing inter-sensing timing parameters that can be used to determine sensor spacing and verification of the repeatability of the AE sensors throughout the testing.

## 4.7 MEASUREMENT SENSITIVITY

Acoustic emissions from electrical discharges are usually monitored by devices that measure sound pressure levels, time varying rarefactions and condensations of the medium through which it travels and are expressed in decibels (dB). Measurements within the frequency range of human hearing from 15 Hz to 20 kHz are classified as sonic, while frequencies less than 15 Hz are termed as infrasonic and the higher frequencies above 20 kHz are called ultrasonic. Acoustic emissions from electrical discharges belong to ultrasonic range. Sensitivity of human ear for sound is compared in Table 4.1 with some of the other hearing devices.

**Table 4.1.** Comparison of sensitivity of hearing devices [20]

| Device | Frequency range | Sensitivity |
|---|---|---|
| Human ear | 15 Hz – 20 kHz | 40 pC air discharges at a distance of 1m |
| Stethoscope and listening tube | 15 Hz – 20 kHz | 10 pC |
| Microphone | 40 kHz | 5 pC |
| Parabolic microphone | 40 kHz | 20 dB with a 10° directional beam width |

## 4.8 AN OVERVIEW OF AE SYSTEM

Essential features of a faithful AE system shall include, among others, the capability to perform the following tasks:

i. Detect acoustic emissions
ii. Locate the source of AE emission
iii. Provide means for discriminating between signals of interest and noise

AE instruments currently available vary widely in form and sophistication of functions they perform. Majority of the AE systems are designed to comply with either ASME code or ASTM standards [ 67].

The developments in AE technology coincided with those of personal computers and hence are almost always controlled by custom built generic software. Also, data acquisition and storage modules are integral with the system so that the complete set up comes over as a stand-alone system. AE systems are available in sizes from 1 channel to over 100 channels, depending on the size and complexity of the structure to be tested. The stream of event description is directed through a central processor so as to coordinate the several complex tasks of data storage, display, and operator communications. In larger systems, these tasks are assigned to several smaller processors so as to improve system reliability and confine malfunctions to smaller and identifiable compartments. In many systems, the set of numbers representing signal features and other information is stored in disks and provides unlimited post-measurement analytical capability. Complete data storage is a vital aspect of applied AE technology. It reduces dependence on the on-site operator and permits installation of effective data mining software.

The acoustic signals from the electrical apparatus are converted into electrical signals by the sensors, amplified to usable voltage levels by the preamplifiers. After sensing and pre-amplification, the signal is transmitted to the main instrument, where it is further amplified and filtered. Next is the critical step of detecting the signal. This is accomplished with a comparator circuit, which generates a digital output pulse whenever the AE signal exceeds a fixed threshold voltage. A digital description of AE signal is generated by the front-end hardware and is passed on to a computer system, which provides data storage, a variety of graphical displays, and replay for post-test.

For on-site testing of power transformers and other electrical equipment, PC based multi-channel AE systems are generally employed. Most of the present day AE systems are fully digital, multi-channel and computerized systems. Figure 4.4 (a) depicts a view of an AE system and Fig. 4.4(b) a simplified block diagram of a four channel of the AE system.

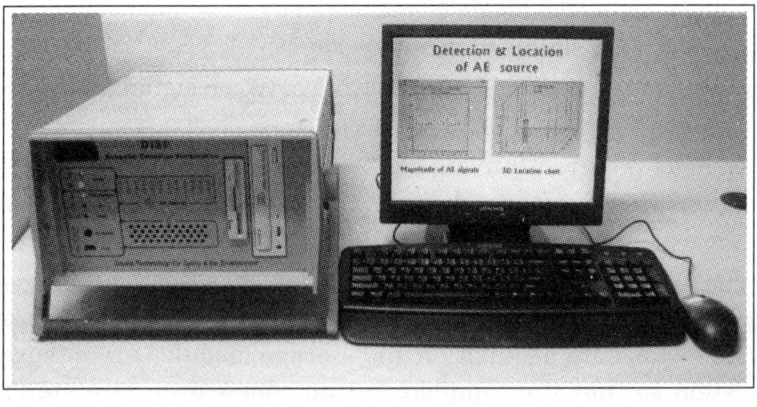


**Fig. 4.4 (a).** A view of an AE system

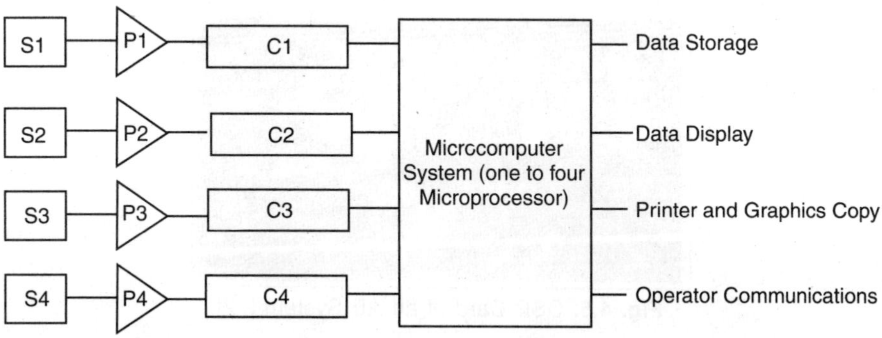


S1 to S4: AE sensors, P: Pre amplifier, C1 to C4: Signal measuring units

**Fig. 4.4(b).** Block diagram of a four-channel Acoustic Emission System [72]

## 4.9 SIGNAL PROCESSING UNIT

The most important part of an AE system is the AE signal processing board which performs parallel digital signal processing (DSP). Some of its important features of commercially available AE systems are as follows:

i. Each board processes data received in the digital form from the different channels, generally.
ii. Peripheral component interconnect (PCI) bus configuration provides AE data transfer rates of up to 132 Mega bits per second to computer.
iii. The A/D converter 16 bits, with a clock rate of 10 MHz.
iv. High pass and low pass filters.
v. High-performance index of 32 bit (in floating point digital signal processing mode).
vi. On board Auto Sensor Test (AST) circuitry which provides software generated and controlled pulse through sensors.
vii. Custom built software for real time measurement, analysis, display and storage of AE data for geometrical defect location (3D).

Figure 4.5 shows a view of DSP card and Fig. 4.6 shows the block diagram of the DSP board. Due to advancement in technology, compact AE systems with channels up to 8, 16, 24 or 56 in a single enclosure with integral PC are commercially available. For example, a 24 channel AE system weighs less than 25 kg with dimensions as 0.50 m long × 0.35 m wide × 0.20 m tall.

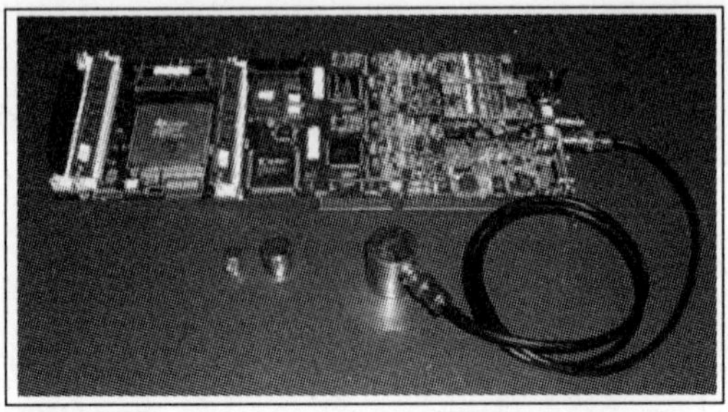

**Fig. 4.5.** DSP Card of an AE System [72]

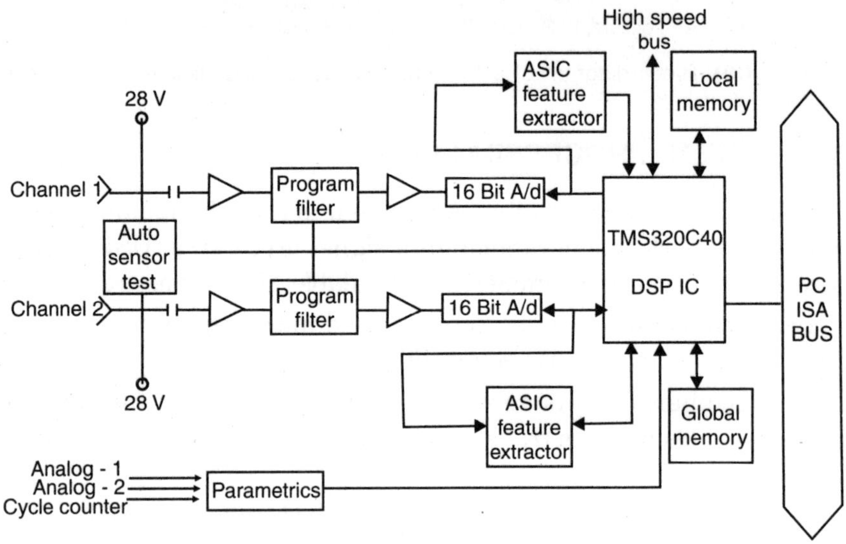


**Fig. 4.6.** DSP Board block diagram [72]

## 4.10 FREQUENCY FILTERING

AE systems provide filtering both in the pre-amplifier and on the signal processing board. The preamplifier uses a pluggable filter module with a specified type and frequency value. This filter can be "high pass", "low pass" or "band pass". The most standard selection is a band-pass filter with 10 kHz to 200 kHz bandwidth for electrical application. However, with the multiple selection filter capability built into the signal processing boards, it is recommended that a wider bandwidth filter is used and that the actual filter selection is made with software control. This allows flexibility in selecting sensors and frequency ranges without requiring any internal hardware or manual changes.

The signal processing boards come with a pluggable filter module, which contain 4 high pass and 4 low pass selections. Combination of high pass and low pass filters provides many different filtering ranges.

These filters can be selected through computer software. The standard filter frequency values are listed below:

| High pass filters | | Low pass filters | |
|---|---|---|---|
| HP-1 | 10 kHz | LP-1 | 100 kHz |
| HP-2 | 20 kHz | LP-2 | 200 kHz |
| HP-3 | 100 kHz | LP-3 | 400 kHz |
| HP-4 | 200 kHz | LP-4 | 1MHz |

## 4.11 SELECTION OF AE SENSORS

The most important factor in AE testing is the selection of appropriate type of sensor. A large variety of sensors are available such as general purpose, wide band, high temperature, miniature, integral type and so on. Depending on the application, suitable type of sensor shall be selected. Important factors influencing the selection are frequency spectrum of acoustic emission, temperature of the test object, shape of the object etc. Some of the commercially available sensors are listed in Table 4.2.

**Table 4.2.** Characteristics of typical commercially available AE sensors [73]

| Type | Sensor model | Operating temp °C | Material | Operating frequency range, kHz | Resonant frequency, kHz | Peak sensiti-vity, ref volts/ m/sec | Direc-tionality, dB |
|---|---|---|---|---|---|---|---|
| Special purpose | Rolling sensor: RSWD | –45 to +100 | Silicone rubber | 200-1000 | 250 | 50 | ±1.5 |
| | Air borne: R2 | –45 to +110 | Al | 15–25 | 20 | -- | –20 dB at 45°C |
| | Under water: R15IU | –45 to +85 | Ceramic | 50 – 200 | 125 | 108 | ±1.5 |
| General purpose | R15 | –65 to +177 | Ceramic | 50 – 200 | 150 | 69 | ±1.5 |
| | R80 | –65 to +177 | Ceramic | 200–1000 | 200 | 61 | ±1.5 |
| Wide band | WD | –65 to +177 | Ceramic | 100–1000 | 125 | 87 | ±1.5 |

AE sensors are available as either single-ended design or differential type. The differential designs provide common mode rejection of unwanted signals in environments of high electromagnetic inference. Also, some of the sensors are integrated to the preamplifier, without the need of a separate pre-amplifier which are very useful for on-site measurements.

Sensors are also available with optional Auto Sensor Test (AST) capability. AST feature is of particular importance for quickly checking the test system before actual measurements are started. The technical details of a typical sensor are shown in the Table 4.3 and frequency bandwidth is shown in Fig. 4.7.

**Table 4.3.** Features of an acoustic sensor [73]

| | |
|---|---|
| Sensor type | DT 15I* |
| Pre amplifier gain | 40 dB |
| Operating frequency | 100 kHz to 400 kHz |
| Resonant frequency | 150 kHz |
| Peak sensitivity | –23 dB |
| Operating temperature | –45°C to +85°C |
| Case material | Stainless steel |
| Face material | Ceramic |
| Sensor drive capability | 10 meters |
| Dimensions, mm | 29 × 31 |

* Physical Acoustics make

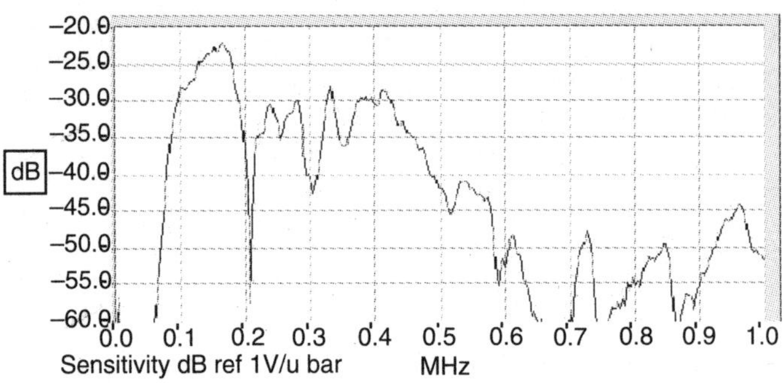


**Fig. 4.7.** Bandwidth of AE Sensor DT15I [73]

## 4.12 CALIBRATION OF AE SYSTEM

AE system is calibrated with the help of commercially available AE calibrator for the required application. A calibrator generates reference waveforms with defined amplitude, frequency and rise time which can be measured with the help of an oscilloscope. The reference waveform is fed to each channel of AE system to compare the amplitude and frequency for different dB levels. A typical calibration result for an AE calibrator and a 16 channel AE system are shown in Table 4.4 and Table 4.5. A view of AE calibrator along with reference waveform is

**Table 4.4.** Calibration of an AE calibrator

| Characteristics of reference wave from AE calibrator | | | Measured values with oscilloscope | |
|---|---|---|---|---|
| Amplitude, dB | Amplitude, Equivalent volt | Frequency | Amplitude, Volt | Frequency |
| 30 | 3.98 mV | 150 | 3.8 mV | 152 |
| 60 | 100 mV | 150 | 105 mV | 151 |
| 80 | 1 V | 150 | 1030 mV | 152 |

**Table 4.5.** Calibration of AE system

Calibrator: AE mode, AE System parameters: PDT = 50µS, HDT = 100µS, HLT = 100µS.

| Amplitude from calibrator | Amplitude measured at different channels, dB | | | | | | | | | | | | | | |
|---|---|---|---|---|---|---|---|---|---|---|---|---|---|---|---|
| | 1 | 2 | 3 | 4 | 5 | 6 | 7 | 8 | 9 | 10 | 11 | 12 | 13 | 15 | 16 |
| 30 dB | 30 | 30 | 31 | 30 | 31 | 30 | 30 | 30 | 29 | 30 | 30 | 31 | 30 | 30 | 31 |
| 60 dB | 59 | 59 | 60 | 59 | 59 | 60 | 60 | 60 | 59 | 60 | 60 | 60 | 60 | 59 | 60 |
| 99 dB | 99 | 99 | 99 | 99 | 99 | 99 | 99 | 99 | 98 | 99 | 99 | 99 | 99 | 98 | 99 |

shown in Fig. 4.8. Measured amplitudes are within a tolerance of ±2 dB, which is acceptable for acoustic PD testing.

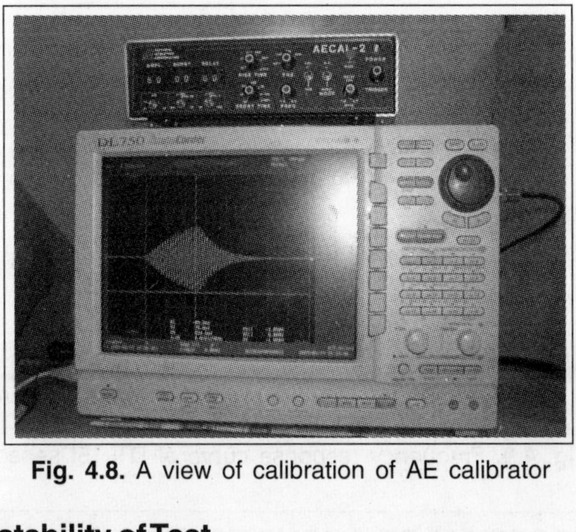


**Fig. 4.8.** A view of calibration of AE calibrator

### 4.12.1 Repeatability of Test

Checking of repeatability/reproducibility of test results is essential for AE systems which are regularly used for on-site tests. Suggested periodicity for repeatability/reproducibility test is bi-annual.

A reference sample is to be identified for periodic calibration of AE system and checking the repeatability of AE PD measurement. For satisfactory reproducibility, measured values shall differ by less than ± 3 dB. Typical results of AE PD test conducted on a reference test sample for two consecutive years are given in Table 4.6.

**Table 4.6.** Periodic measurement of AE PD on a test sample

| Reference test sample | Sensor used | Test voltage | Initial acoustic PD | | Repeat measurement after one year | |
|---|---|---|---|---|---|---|
| | | | January | July | January | July |
| Prototype transformer | DT15I | 20 kV | 75 dB | 76 dB | 74 dB | 76 dB |

### 4.12.2 Verification of Frequency Response of AE Sensors

Verification of frequency response of sensors is necessary to check the condition of sensor and is accomplished by comparison of frequency response curve with the manufacturers calibration chart. For example, frequency response curve of DT 15I sensor measured in the laboratory (Fig. 4.9) is compared with its standard calibration chart (refer to Fig. 4.7) which verifies the performance of sensor as a good match is observed for peak frequency (150 kHz) and frequency band (100 kHz to 400 kHz).

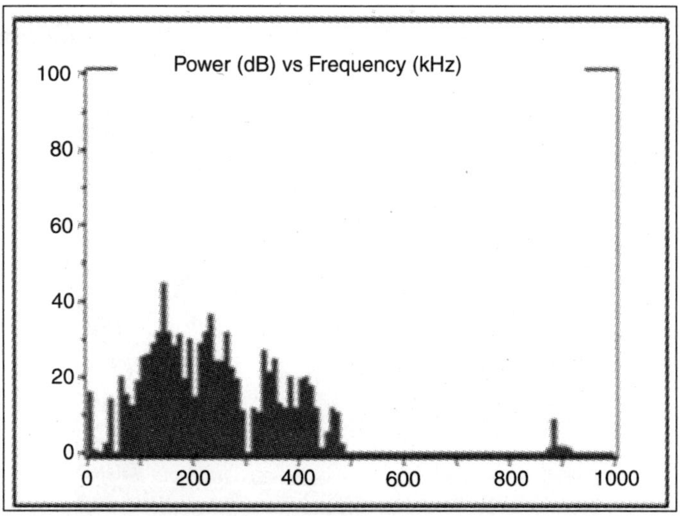


**Fig. 4.9.** Frequency response curve of DT 15I sensor

## 4.13 PROPAGATION OF ACOUSTIC SIGNAL

In the process of detecting partial discharges using acoustic signals measurement it is important to understand the shape of the AE signal received at the sensors and its characteristics will be fundamentally governed by the propagation process. The important aspects of wave propagation in typical AE testing are attenuation and wave velocity. Attenuation is the loss of signal amplitude due to geometric factors and material damping as the wave travels through the material [20, 39].

Attenuation governs detectability at a distance and is therefore an important factor in choosing sensor positions and spacing. AE measurement procedures typically call for attenuation measurements to be made before a test and specify permissible sensor spacing based on these measurements. Wave velocity is an additional factor to be considered when AE technology is used for source location. The acoustic propagation path from the source of discharge to the sensor is specific to the apparatus under test and ranges from the simple case of discharges on overhead lines, where the acoustic wave propagates through air only, to the complicated case of propagation through the complex structures of power equipment such as power transformers.

Acoustic property of a material depends on both the composition and the production process of the material. Absorption of AE signals in solid is a very complicated phenomenon and the absorption coefficient can vary over several decibels. In addition to absorption caused by viscosity and thermal conductivity, losses are also associated with boundaries at in-homogeneities (as in a porous material). This loss mechanism has much in common with scattering of light resulting from variations in index of refraction within an optical material. As signal detection sensitivity is reduced by absorption, it sets limits for the maximum thickness of a test object for a given detection sensitivity of PD.

Acoustic emission during PD activity within a void in solid insulation immersed in a liquid insulating medium in a metal enclosure, has to pass through solid and liquid insulation and then through the thickness of enclosure before reaching the transducers placed on and acoustically coupled to the outer wall. In so doing, the signal reaching the sensor undergoes considerable change, attenuation, distortion and delay, depending on the acoustic properties of the constituent materials which limits the overall efficiency of acoustic PD detection system.

Depending on the direction of propagation, acoustic emissions are classified as longitudinal waves, transverse or shear waves and Raleigh-Lamb waves. The longitudinal wave in a medium consists of alternate compressions and rarefactions of the medium in the direction of propagation. The macroscopic particles of the matter vibrate about their equilibrium positions parallel to the axis of the wave. In transverse or shear waves, the particle motion is perpendicular to the direction of propagation, for example, similar to vibrations in a string. Raleigh-Lamb waves or surface waves are complex form of longitudinal and transverse waves which travel on the surface of the material. The characteristics of AE PD signals as measured are therefore, depend on the signals generated at the source, its location and the path traversed by the signals to reach the sensors.

Gaseous and liquid medium allow propagation of only longitudinal waves. Solids allow both longitudinal waves and transverse waves. Surface waves can only travel in solids. The transverse waves travel at half the velocity of longitudinal waves. Surface waves travel at a velocity of 90 % of transverse waves. For example, for Aluminium, transverse wave velocity is 3080 m/s, longitudinal wave velocity is 6260 m/s and surface velocity is 2800 m/s [20].

### 4.13.1 Characteristic Impedance of Propagation Medium

The characteristic impedance or specific acoustic impedance of a material is the product of the density (ρ in kg/m³) of the material and velocity ($c$ in m/s) of sound in that material and is expressed in ρc in kg/m²/s. Considerable difference exists between the characteristic impedances of solids, liquids and gases. For example, the respective values for steel, water and air are 39.4, 1.48 and 0.0004*10⁶ kg/m²/s.

### 4.13.2 Reflection Coefficient of Propagation Medium

At the interface of two dissimilar media, or a discontinuity, acoustic waves are reflected and refracted. The reflection coefficient at the interface is given by:

$$R_0 = \frac{(\rho_1 c_1 - \rho_2 c_2)}{(\rho_1 c_1 + \rho_2 c_2)} \tag{4.1}$$

where, ρ and $c$ are respectively density and velocity of sound in the medium. A compilation of Ogihara [20] of reflection coefficient for some of the materials is given in Table 4.7 and acoustic properties of some of the engineering materials are listed in Table 4.8.

**Table 4.7.** Acoustic reflection coefficient at interfaces of materials [20]

| Material | $V_L$ m/s ($10^3$) | ρ kg/m³ ($10^3$) | Reflection coefficient with respect to air (Normal Plane Wave Incidence) | | | | | | | | |
|---|---|---|---|---|---|---|---|---|---|---|---|
| | | | Air | T/F oil | Water | Polystyrene | Quartz Glass | Brass | Cu | Steel | Al |
| Al | 6.35 | 2.69 | 100 | 74 | 72 | 51 | 0.7 | 16 | 16 | 20 | 0 |
| Steel | 5.88 | 7.00 | 100 | 89 | 88 | 78 | 27 | 0.5 | 0.5 | 0 | -- |
| Copper | 4.40 | 8.93 | 100 | 88 | 87 | 75 | 21 | 0 | 0 | -- | -- |
| Brass | 4.64 | 8.54 | 100 | 88 | 87 | 75 | 22 | 0 | -- | -- | -- |
| Quartz Glass | 5.57 | 2.6 | 100 | 70 | 67 | 45 | 0 | -- | -- | -- | -- |
| Poly-styrene | 2.67 | 1.06 | 100 | 14 | 11 | 0 | ---- | -- | -- | -- | -- |
| Water | 1.43 | 1.0 | 100 | 0.30 | 0 | -- | -- | -- | -- | -- | -- |
| T/F oil | 1.39 | 0.8 | 100 | 0 | -- | -- | -- | -- | -- | -- | -- |
| Air | 0.331 | 0.001 | 0 | -- | -- | -- | -- | -- | -- | -- | -- |

$V_L$ : Longitudinal velocity, Cu: Copper, Al: Aluminium, T/F: Transformer

**Table 4.8.** Acoustic properties of some engineering materials [20]

| Substance | Longitudinal velocity, c (m/s) | Density r (kg/m$^3$) ($10^3$) | Acoustic impedance (kg/m$^2$/s) ($10^6$) | Poisson's ratio, r |
|---|---|---|---|---|
| Aluminium | 6400 | 2.7 | 17.28 | 0.355 |
| Steel | 6000 | 7.8 | 39.39 | 0.25 – 0.33 |
| Brass | 4280 – 4700 | 8.6 | 38.61 | 0.374 |
| Copper | 4700 | 8.9 | 42.00 | 0.26 |
| Quartz Glass | 600 | 2.6 | 15.60 | 0.17 |
| Polyethylene | 2000 | 0.9 | 1.80 | 0.458 |
| Polystyrene | 2.350 | 1.056 | 2.48 | 0.405 |
| Poly-Tetra – Fluoroethylene (Teflon) | 1.35 | 2.1-2.3 | 3.00 | -- |
| Pressboard impregnated with mineral oil | 2000 | 0.9 – 1.3 | 2.2 | -- |
| Water | 1.484 | 0.999 | 1.48 | -- |
| Mineral oil | 1.39 | 0.90 | 1.25 | -- |
| Caster oil | 1500 | 0.95 | 1.425 | -- |
| Air, 0°C, 1 atm | 331 | 0.001293 | 0.000428 | -- |
| Nitrogen | 333 | 0.001250 | 0.000416 | -- |
| Oxygen | 315 | 0.001429 | 0.000450 | -- |
| Carbon dioxide | 260 | 0.001977 | 0.000510 | -- |
| Hydrogen | 1286 | 0.0000899 | 0.000116 | -- |
| $SF_6$, 25°C , 1 atm, | 140 | 0.006602 | 0.000924 | -- |

## 4.14 ATTENUATION

Acoustic waves are attenuated along their path of propagation due to wave diffusion losses, collisions, losses at interfaces and losses due to molecular absorption, viscosity, and heat conduction. Diffusion losses are more in gases and liquids, whereas, in solids, attenuation will be predominantly due to heat generation. Attenuation depends on the medium of propagation, temperature, humidity and frequency. A compilation of attenuation of waves in certain solids, liquids and gases is given in Table 4.9 [20].

**Table 4.9.** Attenuation of acoustic signals in different media [20]

| Medium | Frequency | Longitudinal wave Attenuation, dB/m |
|---|---|---|
| Air | 50 kHz | 0.98 |
| $SF_6$ | 40 kHz | 26.00 |
| Distilled water | 50 kHz | 0.00027 |
| Sea water | 50 kHz | 0.005 |
| Transformer oil | 3 MHz | 37 |
| Aluminium | 10 MHz | 9.00 |
| Steel | 3 MHz | 21.5 |
| Polystyrene | 2.5 MHz | 100 |

## 4.14.1 Estimation of Attenuation

Attenuation of AE signals can be estimated with reference to PD source of fixed characteristics, particularly, the amplitude. AE signals of amplitude 100 dB generated from a calibrator are pulsed at a reference sensor and amplitude of AE signals at the receiving sensor is measured. The simplest arrangement of sensors for estimating the attenuation is shown in Fig. 4.10. The two sensors, pulser and receiver, are so positioned that AE signals propagate through individual or combination of different insulation media before reaching the receiver. Attenuation of AE signal is estimated by the difference in AE signal amplitude between the pulser and the receiver. Typical values estimated for different insulating media and the transformer container material are depicted in Fig. 4.11[64]. As can be seen from Fig. 4.11, oil medium offers minimum attenuation for the propagation of AE signals.

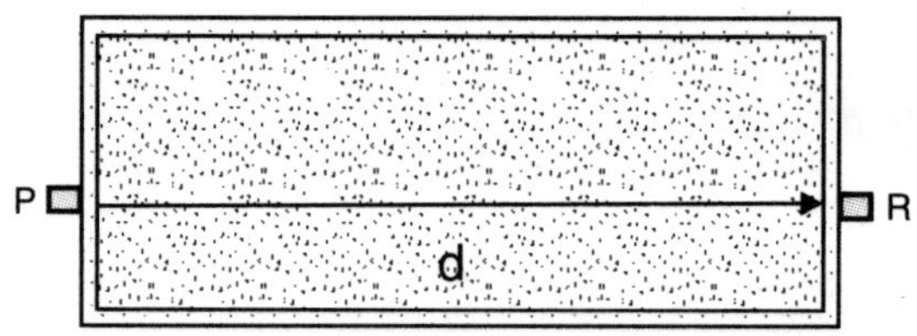


P: Pulsar, R: Receiver

**Fig. 4.10.** Arrangement for estimating the attenuation of AE signals

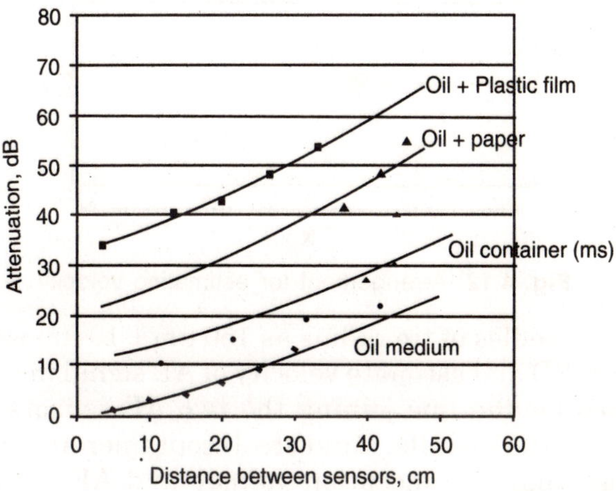


**Fig. 4.11.** Signal attenuation in different insulating media

## 4.15 PROPAGATION VELOCITY

Velocity of AE signal depends on the medium in which it travels and is influenced by several parameters like, propagation length, propagation direction, temperature, etc. One of the inputs required for estimating the geometrical location of AE source is the velocity of AE signal in the media under consideration. Therefore, it is essential to estimate the velocity of the AE signals in the required medium by simulating the AE source as given in the following.

### 4.15.1 Estimation of Wave Velocity

A PD source emits AE signals in all directions. As soon as the wave reaches any one of the sensors, the AE system resets and the counter starts. As soon as the wave reaches the next sensor, the time counter stops and the time delay is registered. Knowing the distance between the sensors and the time delay, velocity, $V$, of AE signals can be calculated from the formula,

$$V = \frac{x - 2d}{\Delta t} \tag{4.2}$$

where, $\Delta t$ is the time delay in receiving AE signals at the second sensor with reference to the first sensor and $d$ is the distance between the PD source and the reference sensor and $x$ is the distance between the sensors, as shown in Fig. 4.12.

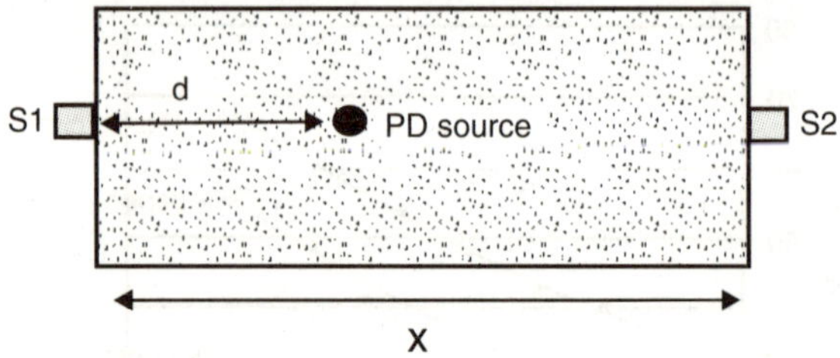


**Fig. 4.12.** Arrangement for estimating velocity

Stainless steel needles of tip radius as 150 μm ± 10 μm were employed to simulate AE PD and estimate velocity of AE signal in oil. The needle was positioned on the line joining the two AE sensors fixed on the opposite sides of the metallic (mild steel) container and energized to a voltage higher than PD inception voltage and AE PD signals were acquired. Time delay ($\Delta t$) and the distance ($d$) from the reference sensor were obtained from the AE system. Time delays recorded for 5 trials at 10 kV are tabulated in Table 4.10 along with the estimated wave velocity. It can be seen that the mean velocity of AE signals is ~ 1250 m/s in the transformer oil used for the study. Results of similar measurements performed on paper - oil system is shown in Table 4.11 where the mean velocity is around 840 m/s.

**Table 4.10.** Wave velocity in transformer oil [64]

| Trial | AE system data | | Wave velocity, V, (m/s) |
|---|---|---|---|
| | d* (m) ($10^{-2}$) | Time delay $\Delta t$, (μs) | |
| 1 | 9.09 | 135 | 1246 |
| 2 | 8.92 | 137 | 1252 |
| 3 | 8.92 | 137 | 1252 |
| 4 | 9.00 | 136 | 1250 |
| 5 | 9.00 | 136 | 1250 |

* Actual distance (*d*) from reference sensor was 0.09 m.

**Table 4.11.** Wave velocity in oil impregnated paper (2 mm)

| Trial | AE system data | | Wave velocity, V (m/s) |
|---|---|---|---|
| | d* (m) ($10^{-2}$) | Time delay $\Delta t$, (μs) | |
| 1 | 9.09 | 201 | 837 |
| 2 | 8.92 | 200 | 841 |
| 3 | 8.92 | 200 | 841 |
| 4 | 9.00 | 201 | 837 |
| 5 | 9.00 | 200 | 841 |

* Actual distance (*d*) from reference sensor was 0.09 m.

## 4.16 DETECTION OF PD

The process of acquiring measurable signals due to a pre-electrical breakdown being indirect there is always an element of doubt as to the comparison of seemingly similar effects produced by the same cause. As often happens, this inconsistency is in part due to the measuring equipment and the way in which the electrical pulses are translated to equivalent acoustic (mechanical) output. Acoustic sensing is essentially a measure of rarefactions and condensations of the media through which the wave traverses and its interaction with the diaphragm of the acoustic sensors. Normally, piezo-electric processes are invoked in which a time varying mechanical stress/strain signal is captured and converted to electrical output. As can be seen, the signal originally was due to partial discharge (electrical) then converted to a mechanical equivalent quantity and reconverted to an electrical signal for display and quantification. The signal therefore, undergoes attenuation, distortion and mismatch at every stage of conversion and hence, *perse* a straight forward comparison can hardly be made as shall be explained later. Using laboratory controlled experimental data a plausible working relationship with measurable and measured quantities can be established on a statistical scale.

### 4.16.1 Comparison of Acoustic and Electrical Methods

Simultaneous measurement of PD using electrical and acoustic methods on a common measuring platform serves two purposes; simultaneous logging both measurements and providing a ready reckoner for correlating electrical and acoustic measurements. In order to be able to formalize the location procedure on site, it is essential to acquire data on laboratory controlled specimen to establish repeatable benchmarks. Keeping this in view, measurement of both electrical and acoustic PD was carried out simultaneously, on a scale down prototype transformer. The AE sensors were appropriately mounted at suitable positions on the outer surface of transformer tank as shown in Fig. 4.13(a).

(a)

Electrical PD was measured in pC and acoustic PD in dB. Typical waveform of AE signal recorded by AE system and electrical pulses displayed on ERA discharge detector due to PD activity are shown in Figs. 4.13 (b) and (c), respectively. Sensitivity for electrical method was 100 pC as per IEC 60076-3: 2000 [68] and threshold for acoustic method was set at 35 dB.

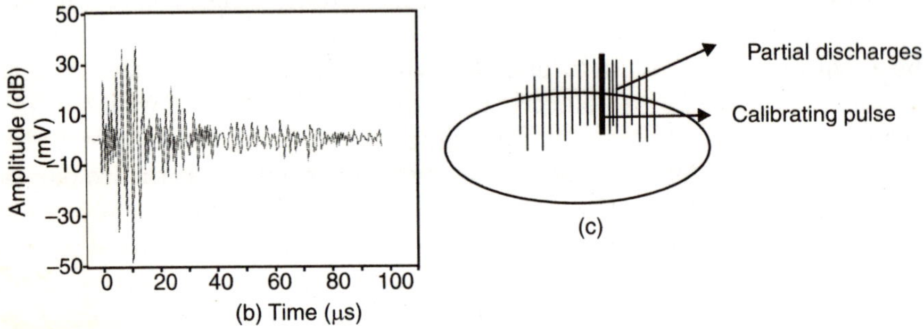


**Fig. 4.13.** Simultaneous measurement of acoustic PD and electric PD, **(a)** Experimental set up, **(b)** Typical acoustic PD waveform, **(c)** Typical electrical PD pulses

A typical set of results of simultaneous measurement of electrical PD and acoustic PD as a function of applied voltage is given in Table 4.12. The data presented herein demonstrates the appropriateness of the choice of test method and that of AE sensors used in the measurements.

**Table 4.12.** Acoustic & Electrical PD as a function of applied voltage

| Applied voltage, kV | Amplitude of electrical PD (max), pC | Magnitude of acoustic PD (max), dB |
|---|---|---|
| 10 | <100 | < 35 |
| 12 | 550 | 48 |
| 14 | 800 | 50 |
| 16 | 1200 | 82 |
| 18 | 1500 | 85 |
| 20 | 2000 | 89 |

## 4.17 LOCATION OF PD SOURCES

The ability to locate AE sources is one of the most important features of the multi-channel AE detection system. This aspect of AE technique is the main reason for gaining importance over conventional electrical technique. As acoustic PD is a transient process, measurement of time arrival of signals at different sensors is necessary to locate the source. AE source is located by measuring the time delay between two or more AE sensors in an array. Location of AE source requires the knowledge of the velocity of AE signals in the medium under study. Based on the input given by the operator, built-in software in the AE system locates the source directly. Among the various options for source location, like, linear, triangular, rectangular and three dimensional (3D) location, suitable one can be selected depending on the placement of AE sensors on the test object. In addition to the location chart, sequence of sensors receiving AE signals, the highest amplitude of AE signal received by individual sensor, time delay and estimated (x, y, z) co-ordinates are also generally provided by the AE system. An extract of such information is shown in Fig. 4.14[64].

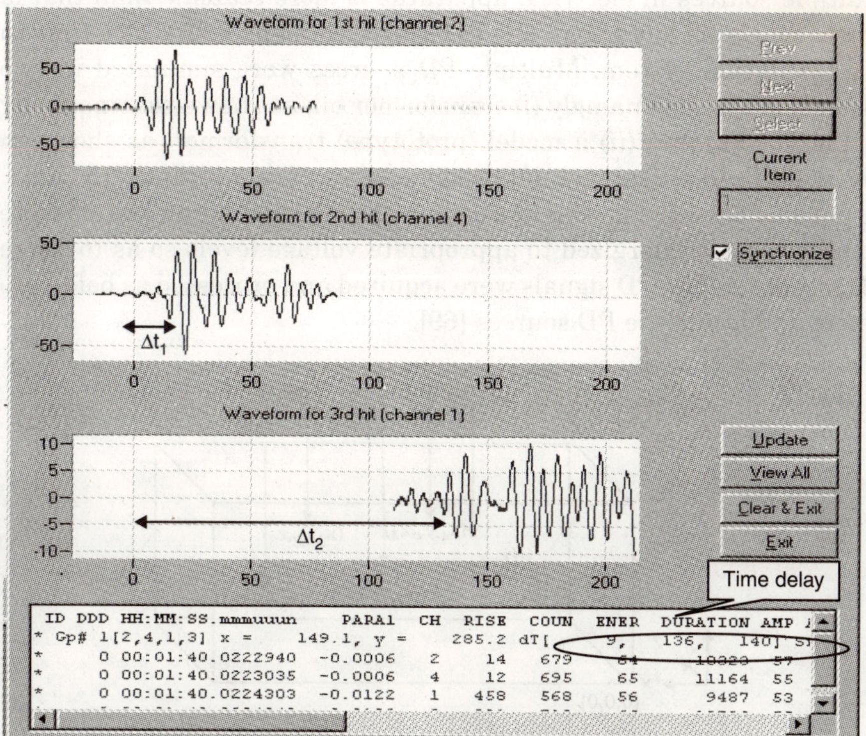


**Fig. 4.14.** Waveforms recorded along with time delay for estimating PD source coordinates

Insulation system of electrical apparatus generally comprises of variety of materials. AE signals generated in insulation system, therefore, have to traverse through different media before they reach the container of the apparatus. Also attenuation, wave velocity and other properties of the AE signals in heterogeneous medium depend on the direction in which they travel before reaching the sensor. Attenuation and wave velocity are the two important aspects to be considered for localization of AE source.

### 4.17.1 Laboratory Study – Detection and Location of Simulated PD Sources

To establish a benchmark, some of the important characteristics of AE signal such as amplitude, attenuation, wave velocity, frequency content, waveform and the like, PD source was simulated with known Cartesian coordinates, $(x, y, z)$. Simulation of PD sources at known locations helps in verification of location results also. Admittedly, the localization of multiple sources in electrical apparatus is more realistic than that of single source as more than one PD source could be active at a time in the insulation system. Multiple PD sources were simulated in two insulation systems namely (*i*) transformer oil, as shown schematically in Fig. 4.15(a) and (*ii*) a model (prototype) transformer, as shown in Fig. 4.16(a). Identical stainless steel needles of tip radius as 150 μm ± 10 μm, positioned at known locations within the insulation system were simultaneously energized to appropriate voltage levels so as to cause PD inception. The PD signals were acquired and processed as before, to detect and locate the PD sources [69].

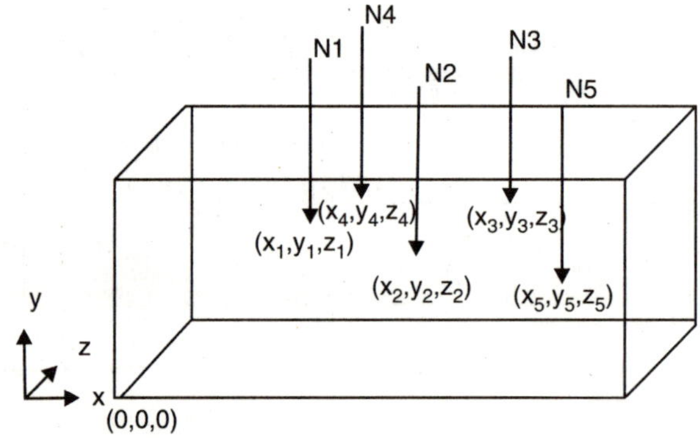


**Fig. 4.15(a).** Arrangement for simulation of PD sources in oil

Figure 4.15(b) shows the magnitudes of acoustic PD occurring at 5 different positions at 18 kV in transformer oil and Fig. 4.15(c) depicts the corresponding 3D location of PD sources. The results of source location so derived are summarized in Table 4.13 along with estimated

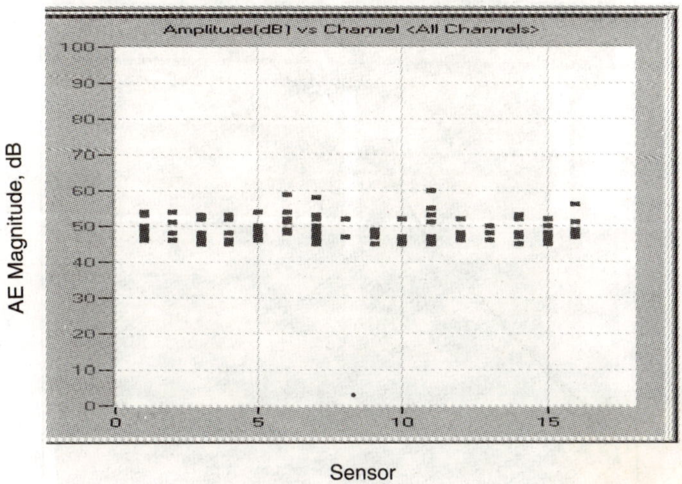


**Fig. 4.15(b).** Acoustic PD due to multiple defects in transformer oil

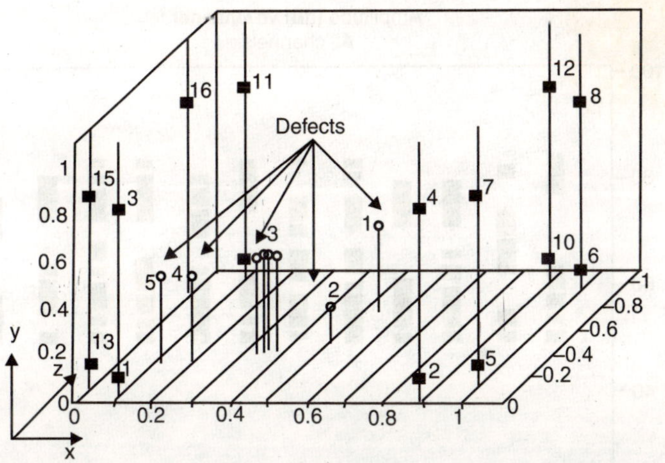


**Fig. 4.15(c).** Location graph of multiple defects in transformer oil

errors for individual coordinates. Similar information for model transformer are depicted in Figs. 4.16(b), 4.16(c) and Table 4.14. The coordinates of PD sources estimated in the above experiments compare well with actual values with an accuracy of ± 9%. While locating defects in a power transformer at site, accuracy to this level cannot be achieved as several factors contribute to error, like inaccuracy due to improper selection of wave velocity, presence of noise and so on. This aspect will be dealt in details in further sections.

**Fig. 4.16(a).** A view of winding of the model transformer

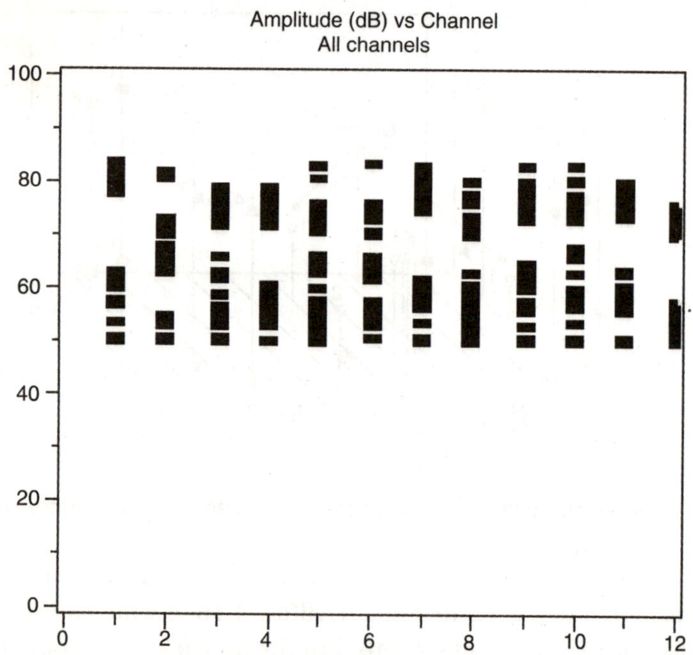


**Fig. 4.16(b).** Acoustic PD due to multiple defects in a model transformer

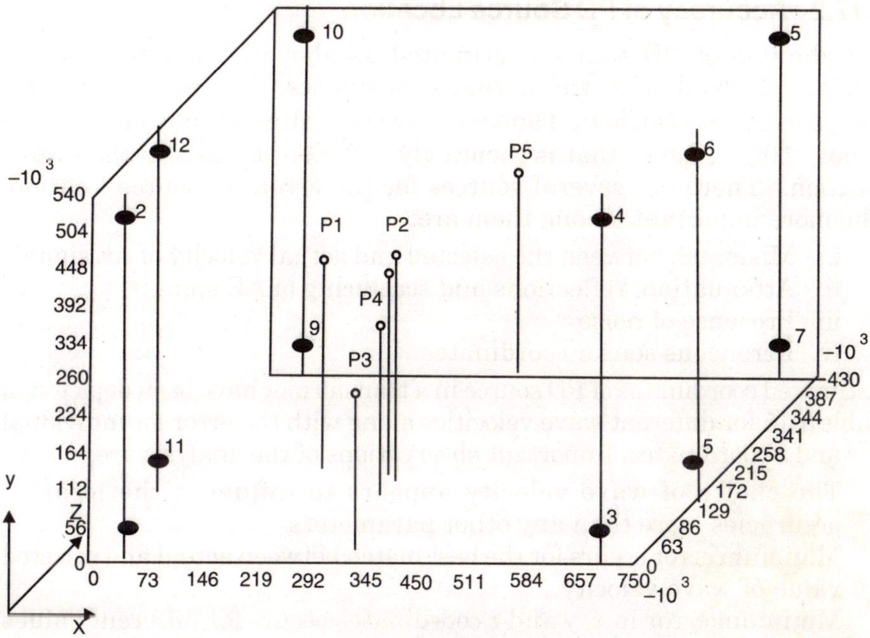


P1, P2, P3, P4 and P5 Defects

**Fig. 4.16(c).** Location graph of multiple defects in a model transformer

**Table 4.13.** Localization of multiple PD sources in transformer oil

| Defect | Actual coordinates, m ($10^{-2}$) | | | Estimated coordinates, m ($10^{-2}$) | | | Error, % | | |
|---|---|---|---|---|---|---|---|---|---|
| | x | y | z | x | y | z | x | y | z |
| 1 | 12 | 39 | 30 | 11.7 | 38.7 | 30.6 | 2.2 | 0.8 | –2.1 |
| 2 | 35 | 39 | 39 | 36.2 | 40.8 | 40.9 | –3.5 | –4.7 | –4.9 |
| 3 | 48 | 12 | 47 | 50.4 | 11.5 | 48.4 | –5.0 | 4.6 | –2.9 |
| 4 | 54 | 37 | 75 | 53.0 | 36.9 | 76.3 | 1.8 | 0.3 | –1.7 |
| 5 | 34 | 41 | 45 | 34.5 | 42.2 | 44.6 | –1.6 | –2.8 | 0.9 |

**Table 4.14.** Localization of multiple PD sources in model transformer

| Defect | Actual coordinates, m ($10^{-2}$) | | | Estimated coordinates, m ($10^{-2}$) | | | Error, % | | |
|---|---|---|---|---|---|---|---|---|---|
| | x | y | z | x | y | z | x | y | z |
| p1 | 30 | 35 | 15 | 30.2 | 35.5 | 14.9 | 0.7 | 1.4 | –0.7 |
| p2 | 28 | 35 | 20 | 29.4 | 34.4 | 19.2 | 5.0 | –1.7 | –4 |
| p3 | 20 | 33 | 20 | 19.3 | 33.7 | 21.6 | –3.5 | 2.1 | 8 |
| p4 | 32 | 22 | 3 | 33.8 | 23.9 | 2.8 | 5.6 | 8.6 | 6.6 |
| p5 | 32 | 32 | 42 | 33.3 | 33.3 | 42.6 | 4.1 | 4.1 | 1.4 |

### 4.17.2 Accuracy of PD Source Location

Coordinates of PD sources estimated as above appear to compare reasonably well with the actual coordinates. A safe accuracy for comparison between actual and measured coordinates may be stated as about ±10%, a figure that is seemingly sufficient to discern the source location. There are several sources for the errors in source location. The more important among them are:

i. Mismatch between the selected and actual velocity of AE signals
ii. Attenuation, reflections and scattering of AE signals
iii. Presence of noise
iv. Erroneous sensor coordinates

Estimated coordinates of PD source in a transformer have been depicted in Table 4.15 for different wave velocities along with the error for individual *x*, *y* and *z* coordinates. Important observations of the analysis are:

- The choice of wave velocity appears to influence the location accuracies more than any other parameters.
- Minimum error occurs for the best match between actual and selected value of wave velocity.
- Minimum error in *x*, *y* and *z* coordinates occurs for different values of velocities, indicating the influence of direction of propagation of AE signals and heterogeneous nature of the propagation medium. Consequently, selection of a single value for wave velocity may not be correct for heterogeneous insulation system.
- Therefore it is essential to estimate the velocity for the medium in which source location is required. In the transformer under consideration, the value lies in the range of 1200 m/s to 1500 m/s.

**Table 4.15.** Precision of source location

| Velocity (m/s) | (x,y,z) Coordinates ( m $10^{-2}$) | Error, % | | |
|---|---|---|---|---|
| | | X | Y | Z |
| 800 | (50.9,44.1, 60.7) | –0.9 | 2.5 | –8.8 |
| 900 | (50.9,43.7, 62.1) | –0.9 | 1.6 | –6.7 |
| 1000 | (51.0, 43.6, 63.5) | –0.7 | 1.3 | –4.6 |
| 1100 | (51.1,43.5, 64.7) | –0.5 | 1.1 | –2.8 |
| 1200 | (51.2, 43.2, 66.0) | –0.3 | 0.4 | –0.9 |
| 1300 | (51.3, 42.9, 67.4) | –0.1 | –0.2 | 1.2 |
| 1400 | (51.4, 42.6, 68.7) | 0 | –0.9 | 3.1 |
| 1500 | (51.4, 42.4, 70.2) | 0 | –1.4 | 5.4 |
| 1600 | (51.5, 42.3, 71.5) | 0.1 | –1.6 | 7.3 |
| 1700 | (51.5, 42.2, 72.9) | 0.1 | –1.8 | 9.4 |
| 1800 | (51.3, 41.8, 74.1) | –0.1 | –2.7 | 11.2 |

Source (x,y,z) : (51.4, 43, 66.6) (m $10^{-2}$)

- Presence of noise from different sources influences the accuracy of location. Elimination of noise enhances the accuracy to large extent, which requires precise knowledge of the characteristics of AE signals from different sources, as illustrated later.
- Source coordinates are estimated by the AE system with reference to sensor coordinates as fed by the operator. Site engineers therefore have to take necessary precautions for accurate measurement of sensor coordinates.

## 4.18 NOISE DURING AE MEASUREMENT

Precaution against interfering noise is an integral part of AE technology. Enormous progress has been made since the early days when students worked at night, using specially constructed underground laboratories to avoid disruption of their experiments by street traffic and people moving nearby. With current technology, many tests can be performed without special measures and a wide range of technologies have been developed to make AE inspection applicable in noisy environments.

A basic starting point is the selection of an appropriate frequency range for AE monitoring. The acoustic noise background is highest at low frequencies. The 100 kHz to 400 kHz range has proved suitable for most of AE testing on power equipment.

Noise problems can be addressed in many ways. First, it may be possible to stop the noise at the source. Second, because noise sources often give characteristically different waveforms, they can often be separated from true acoustic emissions by computer inspection of the measured signal characteristics. This can be accomplished immediately after measurement (front-end filtering), during the display process (graphical filtering), or after the test by replaying the data through a post-test filtering program or advanced waveform analysis package.

Acoustic sources of different types produce acoustic emissions with different characteristics, which help in discrimination of PD source from noise. Important features, which can be employed for noise analysis are amplitude, waveform and frequency profile. Waveform analysis based on FFT, as described in following section, will also help in discriminating PD from noise.

## 4.19 SIGNATURE ANALYSIS OF ACOUSTIC EMISSION SOURCES

During field testing of power transformers, in addition to PD signals, AE data acquired may contain extraneous acoustic signals from sources other than those due to electrical effects, like thermal, mechanical (steady state and time varying structural vibrations) and environmental. Field data acquired from power transformers in service might include one or more of these parasitic disturbances. Filtering techniques of different complexities are often required to discriminate and eliminate the spurious signals. Here again, simulated experiments under laboratory conditions wherein chosen disturbances can be used to observe the real signals contaminated with noise. As applied to power transformers, the parasitic sources can be classified as:

i. Electrical sources – PD, arcing, electrostatic discharges, switching transients
ii. Thermal source – hot spots
iii. Mechanical sources – circulating pump operation, tap changer, loose components, core vibration, automobile noise
iv. Natural causes due to rain, wind

Well known Integral Transform methods involving signal decomposition and subsequent Fourier (convolution, de-convolution) analysis are indicated.

Signature of AE signals recorded during simulation of PD are compared with those from other defects (simulated) like arcing and core vibration, based on the Fast Fourier Transforms (FFT). Typical waveforms recorded during simulated experiments namely, (i) PD in transformer oil, (ii) arcing in transformer oil and (iii) core vibration are shown in Figs. 4.17(a), 4.18(a) and 4.19(a), respectively and their FFT are shown in Figs. 4.17 (b), 4.18 (b) and 4.19 (b), respectively. The pictures are self explanatory. FFT gives the frequency profile of these defects as summarized in Table 4.16.

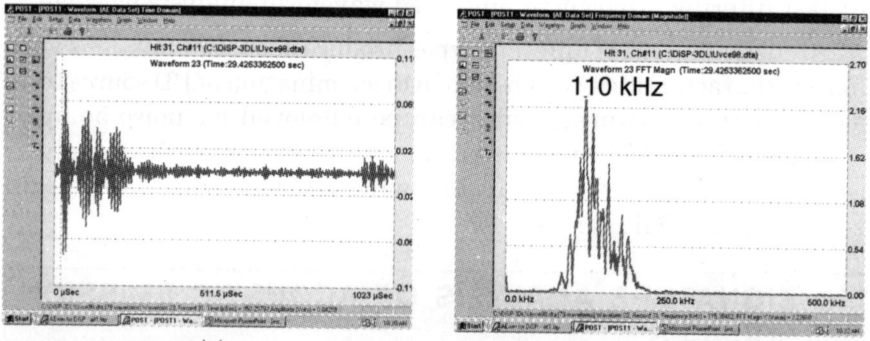


(a) (b)

**Fig. 4.17.** Acoustic signal due to PD in transformer oil **(a)** Waveform, **(b)** FFT of acoustic PD signal

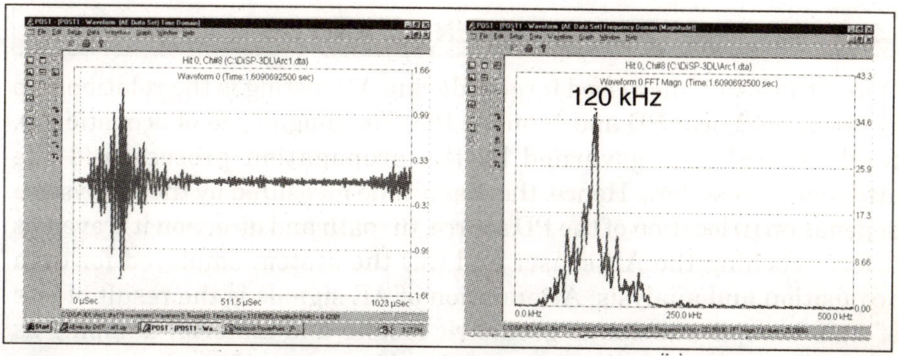


(a) (b)

**Fig. 4.18.** Acoustic signal due to arcing **(a)** Waveform, **(b)** FFT of arc waveform

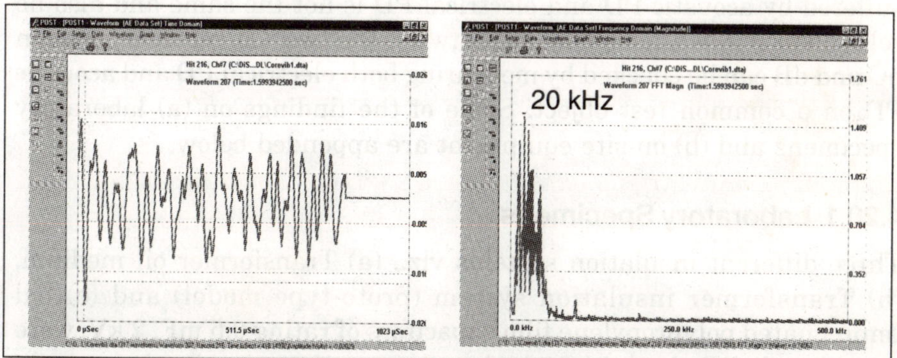


**Fig. 4.19.** Acoustic signal due to core vibration **(a)** Waveform, **(b)** FFT of core vibration

**Table 4.16.** Frequency analysis of defects in power transformers

| Type of defect | Frequency range, (kHz) | Peak frequency (kHz) |
|---|---|---|
| PD / Corona | 100 – 250 | 110 |
| Arcing | 70 – 200 | 120 |
| Core vibration | 5 – 40 | 20 |

Frequency content of AE signals due to arcing are almost similar to PD. However, the core vibration gives AE signals in the frequency range of 20 kHz to 50 kHz, which is low as compared to PD and arcing. Therefore, by selecting suitable frequency range for filtering, unwanted signals can be minimized. For example, by selecting suitable filter range for frequency, i.e., more than 50 kHz, noise due to core vibration can be eliminated.

## 4.20 RELATIONSHIP BETWEEN pC AND dB

One of the issues discussed largely during AE testing is the relationship between electrical PD and acoustic PD. The magnitude of acoustic PD, fundamentally, is governed by the propagation process and the attenuation it suffers. Hence, the magnitude measured by the AE system depends on (i) location of the PD source, (ii) path and direction it traverses before reaching the AE sensor and (iii) the system employed for data acquisition and analysis. Attenuation of AE signals is the result of loss of signal amplitude due to geometric factors and material damping as the wave travels through the material and hence specific to the apparatus under test. Similarly, electrical PD signals also suffer attenuation specific to the equipment under consideration. The extent of attenuation suffered by acoustic PD and electrical PD is not the same and also no relationship is available till date between the two. Correlation between pC and dB can be checked by measuring both electrical PD and acoustic PD on a common test object. Some of the findings on (a) laboratory specimens and (b) on-site equipment are appended below.

### 4.20.1 Laboratory Specimens

Three different insulation systems viz., (a) Transformer oil medium, (b) Transformer insulation system (proto-type model) and (c) Oil impregnated polypropylene film capacitors of rating 0.5 mF, 2 kV, were considered. PD, both electrical and acoustic, was simultaneously measured by applying suitable voltage. The relationships observed between pC and dB are depicted in Figs. 4.20(a), 4.20(b) and 4.20(c), respectively for the above systems. Correlation between pC and dB was examined by applying a transformed regression model. The analysis suggests a logarithmic relation of the form

$$y(\text{dB}) = a\ ln\{x(\text{pC})\} + b \tag{4.3}$$

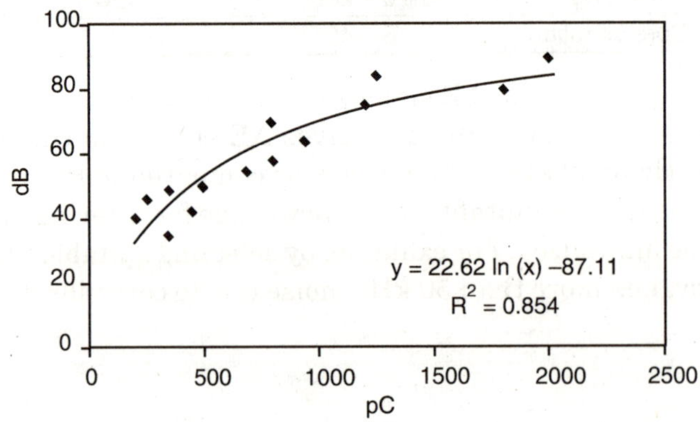


(a)

between magnitude of electrical PD and AE PD, where, '*y*' is the amplitude of the acoustic PD in dB and '*x*' is equivalent electrical PD amplitude in pC.

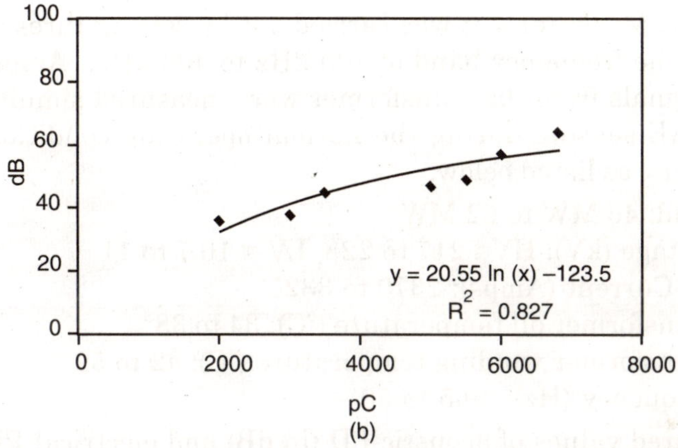


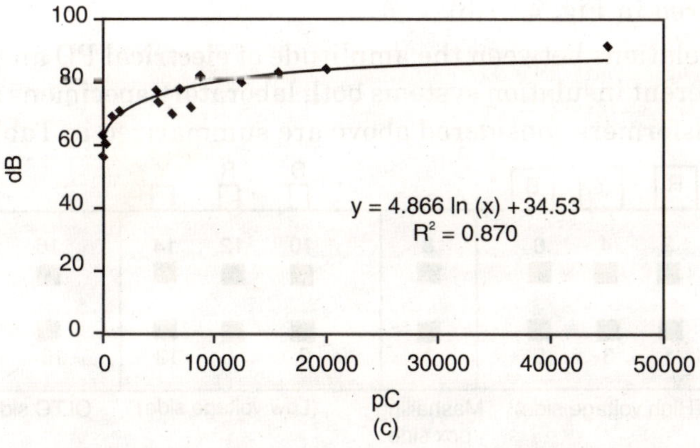


**Fig. 4.20.** Electrical pC vs acoustic dB **(a)** Transformer oil medium, **(b)** Transformer winding and **(c)** Oil impregnated polypropylene film capacitor

### 4.20.2 On-site Equipment -Power Transformers

Correlation between pC and dB has been examined for 25 years old 220 kV, 25 MVA, 1 $\phi$ Generator Transformers operating in a Hydro Power Station. Electrical PD measurement was performed by induced method. For sensing PD, HFCT was incorporated in the neutral lead of the HV winding, as the HV bushing was not provided with the test tap. The measuring system was calibrated in accordance with IEC 60270 by injecting the calibrating pulses at the HV terminal of the HV bushing and sensing them at the neutral lead. At the time of measurement, the noise interference was rather low and the measurement sensitivity was 200 pC. For electrical PD measurement, the transformer was energized by exciting the generator. Acoustic PD measurement was carried out employing a 16-channel AE system by mounting sensors as shown in Fig. 4.21(a). Measurement was carried out by setting threshold at 35 dB and in the frequency band of 100 kHz to 400 kHz. Amplitudes of acoustic signals from the transformer were measured simultaneously at all 16 AE sensors during the normal operating conditions of the transformers, as listed below.

i. Load: 46 MW to 62 MW
ii. Voltage (kV): HV= 217 to 228, LV = 10.7 to 11
iii. LV Current (Amps): 2370 to 3325
iv. Transformer oil temperature (°C): 34 to 38
v. Transformer winding temperature (°C): 42 to 52
vi. Frequency (Hz): 49.5 to 50

The measured values of acoustic PD (in dB) and electrical PD (in pC) are compared in Fig. 4.21(b).

The relations between the amplitude of electrical PD and acoustic PD for different insulation systems both laboratory specimens and EHV grade transformers considered above are summarized in Table 4.17.

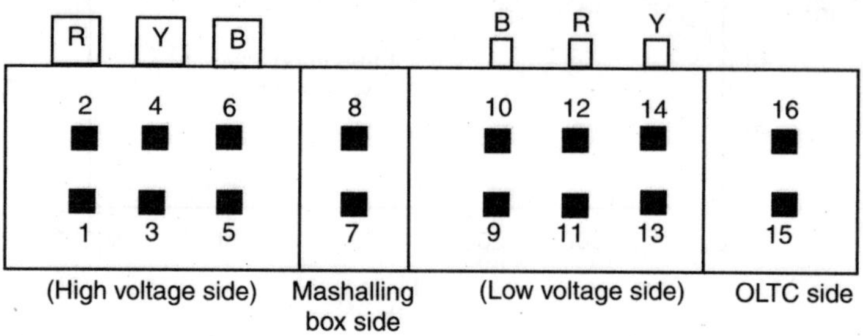


**Fig. 4.21 (a).** Sensor mounting on a power transformer for on-site test

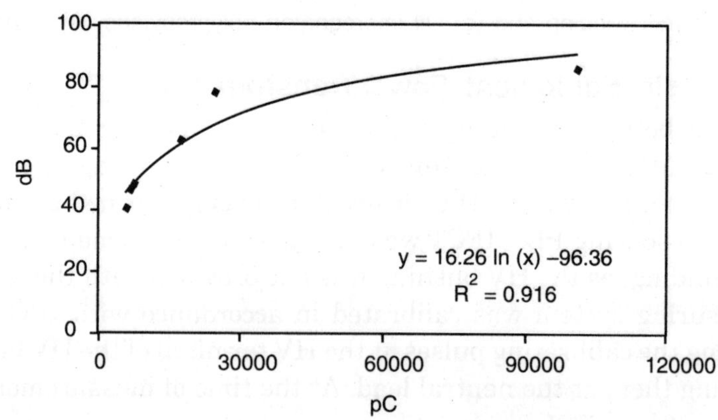


**Fig. 4.21(b).** Electrical pC vs acoustic dB for 220 kV generator transformers

**Table 4.17.** Correlation between pC and dB

| Insulation system | Estimated regression coefficients (Eqn. 4.3) | | |
|---|---|---|---|
| | *a* | *b* | $R^2$ |
| Transformer oil medium | 23 | –90 | 0.85 |
| Impregnated PP film capacitor | 5 | –30 | 0.93 |
| Model transformer | 21 | –123 | 0.82 |
| 220 kV Generator transformer | 16 | –100 | 0.91 |

The logarithmic relationship between magnitude of electrical PD and acoustic PD appears to be reasonably good as the coefficient of determination "$R^2$"is more than 0.8. However, the empirical formulae between pC and dB arrived above are applicable only to the insulation systems tested. The same may not hold good for other equipments due to the following reasons:

i. Propagation process of AE signals may be different in different equipment, due to difference in material, construction and dimension.
ii. The highest amplitude of AE signals as received by the AE sensors is considered as the magnitude of AE PD for the equipment. Sensor close to the most severe AE source receives maximum amplitude AE signals, due to least attenuation. Therefore, the magnitude of AE PD may be different for different equipment depending on the location of the AE source and the sensor position.
iii. Different equipment having different magnitude may show the same electrical pC level. (pC value is not influenced by source location to the extent AE magnitude depends).
iv. AE signals are measured within a narrow range from threshold (say 30 dB) to 100 dB. However, electrical PD signals are measured from 1 pC to few thousands of pC.

In view of the above, the relationship between pC and dB needs to be established for individual equipment.

## 4.21 LIMITS FOR ACOUSTIC PD LEVELS

The most common unit employed for expressing electrical PD amplitude is pico-coulombs (pC). Limits for PD levels have been specified by relevant standards for different equipment depending on the type of equipment, type of insulation, voltage rating, test performed (routine test, type test, test at site), test on new equipment, test on equipment in service, etc. For example, the allowed PD level for power transformer as per IEC 60076-3: 2000 [68] is 100 PC at 1.1 $U_m/\sqrt{3}$ and 300 PC at $U_2/\sqrt{3}$ ($U_m$ and

$U_2$ are defined in Annex D, Tables 2 or 4 of IEC 60076-3: 2000). The safe limits for electrical PD has been arrived based on the huge data bank generated by various organizations throughout the world, as the electrical PD method is in vogue for the last 100 years. The acoustic method is relatively new. Several organizations are actively working on the technique and collection of data has already been started. For instance, Electrical Power Research Institute (EPRI), USA, sponsored project on AE testing has covered more than 500 transformers, so far [35]. The Central Power Research Institute (CPRI), India, has tested so far more than 180 transformers for acoustic PD [48]. Even though, IEEE has introduced trial guidelines for AE based testing of oil filled power transformers [70], limit for AE PD in power transformers is yet to be established, which requires vast database from a large number of power transformers and in-depth understanding of the phenomenon.

AE data collected by the authors from several transformers of rating up to 400 kV have been classified with reference to AE magnitude as shown in Table 4.18. It can be observed that only 2 % of transformers have shown AE magnitude in the range 80 to 90 dB. No transformer has shown PD level more than 90 dB. More than 60 % of the transformers tested have shown magnitude less than 60 dB. AE signals of amplitude less than 60 dB at operating conditions of power transformers appear to be safe as they have been reported to be performing satisfactorily. This preliminary limit is only a starting point and may require revision after augmenting more data on new and aged transformers of different types and ratings. It may also be noted that electrical method specifies the PD level for test voltage, which is generally more than the rated voltage of the system. Electrical PD test voltage will be 1.1 Un to 1.5 Un, depending on test object. However, AE method being on-line and is applied to the equipment at its operating conditions, the test voltage will be generally the rated voltage ± 10%. This aspect needs to be considered while considering the limit for AE PD level.

**Table 4.18.** Acoustic PD in transformers of rating up to 400 kV [65]

| Range of AE magnitude | Number of transformers falling in the range (%) | | |
|---|---|---|---|
| | < 220 kV | 220 kV | 400 kV |
| 100 to 90 dB | 0 | 0 | 0 |
| 89 to 80 dB | 0 | 2 | 2 |
| 79 to 70 dB | 4 | 5 | 9 |
| 69 to 60 dB | 32 | 22 | 16 |
| Below 60 dB | 64 | 70 | 79 |

## 4.22 EQUIPMENT ASSESSMENT BASED ON ACOUSTIC PD TESTING

Once acoustic activity is detected and /or located, the major concern of maintenance engineers is to assess the severity of the fault in order to plan the corrective actions that need to be taken up. Depending on the severity of AE activity, transformers are graded into four categories [35, 48, 71]. Grading system is based on two aspects viz., (a) determination of intensity level of AE activity per cluster (cluster is a well-defined area of the transformer where significant AE activity is occurring as described later) and (b) overall grading. Four different intensity levels have been considered for cluster evaluation. The description of each is given in Table 4.19.

**Table 4.19.** Transformer grading based on acoustic PD test

| Transformer grade | AE test result | Recommendation |
|---|---|---|
| A | No activity detected | No follow up required |
| B | Minor acoustic activity detected. Keep the test data on records for further reference. | No follow up required |
| C | Medium acoustic activity detected, use other diagnostic techniques including DGA to confirm the result. | Requires more frequent measurements |
| D | Considerable acoustic activity registered, immediate action recommended, use other diagnostic techniques including DGA to confirm the result. | Immediate action recommended |

Fixing the intensity levels of acoustic activity for different grades, requires experience and expertise based on large data bank of field test. The Authors' Laboratory is in the process of generating the data bank [48]. This exercise is only a starting point and the proposed grading pattern may require revision after augmenting more data on new and aged transformers of different types and ratings. The main aim of this exercise is to help maintenance personnel in prioritizing the maintenance/ inspection tasks using AE technique.

## 4.23 STATISTICAL ANALYSIS OF ACOUSTIC PD DATA

A cursory look at the PD data may not reveal certain hidden aspects of insulation degradation, unless the data is carefully processed. Given that the PD is a random process, it is expected that nominally identical test objects give vastly differing data, that is, the data is distributed. However, there is a certain order in this chaos which will only be ascertained upon observing the statistical distributions to which the data may conform. Therefore, sound and relevant statistical analysis of data becomes mandatory in identifying of AE signals from different sources and establishing the degree of correlation between different features of AE signals.

Several custom built software packages are commercially available for pattern recognition, signature analysis, quantitative statistics and discriminant analysis.

### 4.23.1 Feature Statistics

Feature statistics give the minimum, maximum, mean, standard deviation, Skewness ($S_k$) and Kurtosis ($k$) of each feature of AE signal. The distributions underlying the AE process is almost always non-normal and infact, does not seem to conform to any known mathematical distribution functions. Often, nonparametric or distribution-free statistical treatment is given to such data [17]. Measures of central tendencies such as, $S_k$ and $k$ appear to be reasonably sensitive 'Feature Statistics' when the distributions are 'Robust'. These parameters are intended to quantify the amount of insulation degradation due to ageing. The Author [17] and certain others [16] have shown on several occasions that the $S_k$ is a particularly interesting parameter of any distribution involving partial discharges. An extensive study of PD magnitude distribution in insulation, described later, has confirmed this finding.

Table 4.20 and Table 4.21, summarize statistical features of AE measurements wherein the simulated data in the laboratory and those obtained from field measurements are compared. The feature recognition, extraction and comparison between two apparently similar sets of data are dependant upon a hidden variable, the measuring environment. For example, laboratory designed simulation experiments and measurements thereof, are much more reliable than site measurements for obvious reasons. The variability including repeatability and reproducibility are often as good as they can possibly be, when simulation experiments are designed in the laboratory. One would therefore expect a higher degree of scatter in all measurements at site. While this has been amply demonstrated by the data in the Tables mentioned above, three parameters, amplitude, signal strength and energy, appear to show a trend to the contrary.

**Table 4.20.** Feature statistics for laboratory data on 60 cm oil gap

| Statistics / AE Feature | Min | Max | Mean | Std. Deviation | $S_k$ | k |
|---|---|---|---|---|---|---|
| **Rise time, μS** | 1 | 1054 | 71.84 | 166.26 | 5.09 | 25.63 |
| **Counts to peak** | 1 | 104 | 7.03 | 12.67 | 6.67 | 48.31 |
| **Duration, μS** | 1 | 43844 | 19602 | 18673 | 0.09 | 1.95 |
| **Avg. Freq, kHz** | 0 | 1000 | 119.75 | 253.62 | 3.09 | 7.84 |
| **Amplitude, dB** | 50 | 82 | 66.97 | 11.66 | –0.09 | 1.71 |
| **Signal strength, pico-volt-sec/count** | 61 | 1.55e+7 | 6.05e+6 | 6.28e+7 | 0.19 | 1.86 |
| **Absolute energy, J/count** | 1.96 | 1.48e+6 | 4.84e+5 | 5.36e+5 | 0.39 | 1.54 |

**Table 4.21.** Feature statistics for the field data*

| Statistical value / AE Feature | Min | Max | Mean | Std. Deviation | $S_k$ | k |
|---|---|---|---|---|---|---|
| **Rise time, μS** | 1 | 4761 | 286.59 | 454.99 | 2.03 | 5.06 |
| **Counts to peak** | 1 | 529 | 5.84 | 23.49 | 8.43 | 91.90 |
| **Duration, μS** | 1 | 1.05e+6 | 89809 | 2.84e+5 | 3.02 | 7.21 |
| **Avg Freq, kHz** | 0 | 1000 | 166.3 | 342.61 | 1.94 | 1.94 |
| **Amplitude, dB** | 35 | 80 | 37.78 | 5.20 | 3.93 | 17.42 |
| **Signal strength, pico-volt-sec/count** | 6.1 | 2.30e+8 | 5.30e+6 | 2.41e+7 | 6.31 | 44.61 |
| **Absolute energy, J/count** | 0.04 | 8.19e+6 | 95236 | 6.53e+5 | 9.41 | 96.63 |

*43.33 MVA, 220 kV Generator Transformer.

Measured amplitude of acoustic signal depends on the severity of the defect and the propagation path between the defect and the sensor and hence on attenuation. The sensor nearest to the acoustic source therefore, receives the highest magnitude. Hence it is not a unique feature of acoustic signal and can not be used for discriminating the defects. Also, signal strength and energy are dependant on amplitude and exhibit the similar statistical behaviour. This analysis is true for both laboratory and field data. Since the acoustic path length, the attenuating medium in the path of the AE, absorption, reflection, refraction and scattering of waves in actual equipment is a very complex aspect. As can be seen from Fig. 4.12, 4.15(a) and 4.16(a), the simulated experiments indicate a rectilinear path with small absorption and scattering of all types will be minimum. In view of this, comparison of parameters which are independent of each other does not seem to be viable.

### 4.23.2 Correlation Matrix

Correlation matrix gives the correlation between different AE features. Correlation matrix of laboratory data and field data are summarized in Table 4.22 and Table 4.23. The AE features which have shown good correlation (correlation coefficient above 0.8) are duration, amplitude,

**Table 4.22.** Correlation matrix for acoustic PD features in transformer oil

| AE Feature | Rise time | Counts to peak | Duration | Amplitude | Frequency | Signal strength | Absolute energy |
|---|---|---|---|---|---|---|---|
| Risetime, µs | 1 | 0.77 | 0.20 | 0.15 | 0.11 | 0.19 | 0.16 |
| Counts to peak | -- | 1 | 0.41 | 0.36 | 0.11 | 0.39 | 0.36 |
| Duration, µs | -- | -- | 1 | 0.97 | 0.23 | 0.99 | 0.99 |
| Amplitude, dB | -- | -- | -- | 1 | 0.35 | 0.95 | 0.92 |
| Avg Freq, kHz | -- | -- | -- | -- | 1 | 0.20 | 0.18 |
| Signal strength, pico volt sec /count | -- | -- | -- | -- | -- | 1 | 0.99 |
| Absolute energy, J/count | -- | -- | -- | -- | -- | -- | 1 |

**Table 4.23.** Correlation matrix for acoustic PD features in a generating transformer

| AE Feature | Rise time | Counts to peak | Duration | Amplitude | Frequency | Signal strength | Absolute energy |
|---|---|---|---|---|---|---|---|
| Risetime, µs | 1 | 0.45 | 0.30 | 0.35 | 0.27 | 0.23 | 0.15 |
| Counts to peak | -- | 1 | 0.61 | 0.71 | 0.05 | 0.70 | 0.57 |
| Duration, µs | -- | -- | 1 | 0.86 | 0.09 | 0.73 | 0.49 |
| Amplitude, dB | -- | -- | -- | 1 | 0.15 | 0.91 | 0.74 |
| Freq, kHz | -- | -- | -- | -- | 1 | 0.04 | 0.02 |
| Signal strength, pico volt sec /count | -- | -- | -- | -- | -- | 1 | 0.93 |
| Absolute energy, J/count | -- | -- | -- | -- | -- | -- | 1 |

signal strength and energy. Comparison of Table 4.22 and Table 4.23 indicates that, as expected, AE features are more correlated for laboratory data than field data.

### 4.23.3 Hierarchical Clustering

Correlation between different AE features can also be examined by hierarchical clustering which groups AE features into pairs and the results are graphically presented in the form of what is called a 'Dendogram' [74]. Each AE feature is initially considered as an independent variable. Then the most correlated descriptions are merged into one, indicated by horizontal line in the dendogram, and a new cluster is formed followed by an update of the correlated matrix.

The merging process continues until all features are merged into a single group. The results of feature correlation hierarchy clustering for the field data on a 220 kV, 25 MVA Generator Transformer is shown in Fig. 4.22. The highest correlated pair of features for the field AE data is formed by energy (feature 4) and signal strength (feature 13). A scatter plot of these two features from a straight line. The other closely correlated features (correlation level above 0.9) are RMS (feature 10), Absolute energy (feature 14), Amplitude (feature 6) and ASL (feature 7).

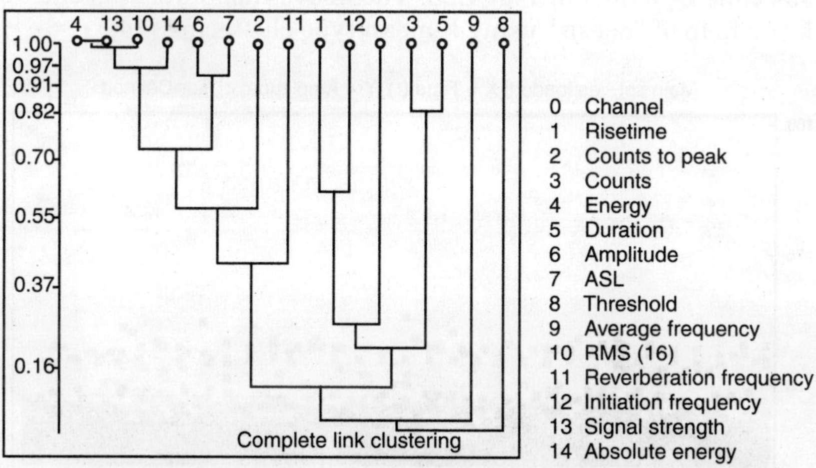


**Fig. 4.22.** Dendogram of field data from a power transformer

The correlation analysis is helpful for selection of proper features. For example, signal strength and energy in the example at Fig. 4.22 have exhibited the highest degree of correlation and one of them can be considered for feature extraction in further tests. The correlation value between amplitude and energy is more than 0.9 indicating noise free field data as correlation lying below 0.7 indicates the presence of noise [74].

### 4.23.4 Acoustic PD Pattern Recognition by Clustering

Classification or grouping or clustering will help to identify the families of patterns or signals with similar characteristics. Clustering is a variant of Kohonen - unsupervised neural network [74] which is obtained by several methods like *k*-means, Max-min distance, Forgy, Isodata, and so on. Among these, *k*-means is a well known algorithm for the minimization of a performance index through an iterative process and is generally attempted for acoustic PD data analysis. Parameters of *k*-mean are (a) Distance type and (b) Initial partitions and (c) Initial clusters. Number of initial partitions and clusters depends on the type of AE data, expected number of different AE sources in the data and more importantly on the understanding of the physical phenomena for that particular test. Generally 100 iterations are made to get the best clustering. It is important to note that quality of clustering results will always depend on the quality of the AE data and inputs given by the operator.

### 4.23.5 Case Study on Statistical Analysis

Cluster analysis for the field data from a 11 / 220 kV, 43.33 MVA, single phase generating transformer at one of the hydropower stations is illustrated below. From the field data, plot of amplitude of acoustic signals versus time is shown in Fig. 4.23. The above field data was clustered with the help of "noesis" using k-mean type clustering [74].

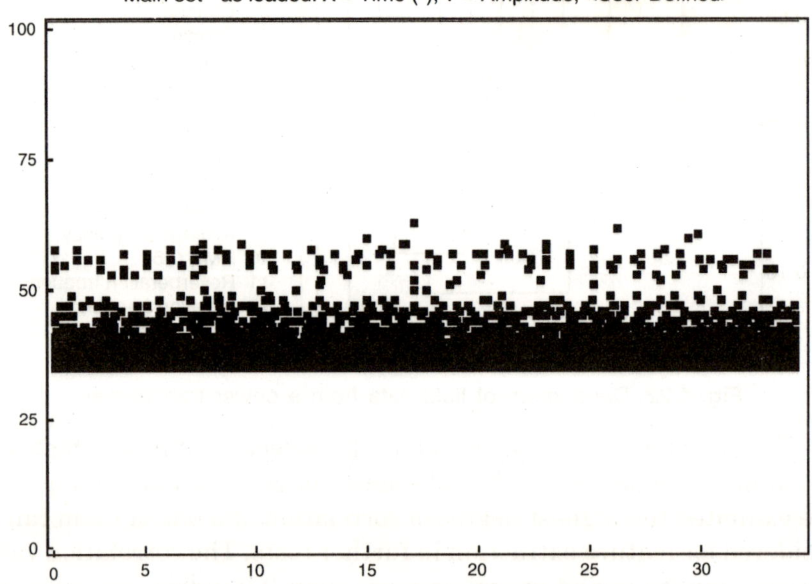


**Fig. 4.23.** A plot of acoustic signal amplitude versus time

The parameters selected for *k*-means clustering were:

Distance type: Euclidean

Initial partitioning: Random

Initial Clusters: 3

Iterations: 100

Clustering of acoustic data for different pairs of acoustic features like amplitude vs. channel, amplitude vs. energy, amplitude vs. average frequency and energy vs. average frequency, is shown in Fig. 4.24 (a), (b), (c) and (d), respectively. As can be seen from these plots, data has been grouped into three clusters, numbered as 0, 1 and 2. Range of values for AE features viz., amplitude, energy, frequency, rise time and duration for the three clusters are given in Table 4.24. As can be seen from Fig. 4.24 and Table 4.24, partition with respect to energy is more discrete as there are no overlaps between clusters. Amplitudes (dB) of different clusters overlap slightly, whereas duration and rise time values overlap considerably with other clusters.

**Table 4.24.** An example of clustering showing ranges of values of acoustic features

| Cluster | Amplitude, dB | Energy, kJ | Frequency, kHz | Duration, ms | Rise time, ms |
|---|---|---|---|---|---|
| 1 | 35 to 52 | 0 - 1.7 | 0 - 1000 | 0 - 400 | 0 - 4 |
| 2 | 45 to 56 | 1.7 to 5 | 20 - 90 | 300- 1000 | 0 - 2 |
| 0 | 52 to 63 | 5 to 12.8 | 55 - 110 | 800- 1000 | 0 - 3 |

As seen in Figs. 4.24 (b), (c) and (d), clustering has resulted in three discrete groups of signals with different energy bands which can be termed as low, medium and high belonging to clusters, cluster 1, cluster 2, cluster 0, respectively. Signals of cluster 1 are received by all the sensors except, 2, 6, 8 and 16. Cluster 1 represents signals of lower amplitude and energy and has a broad frequency range of 10 kHz to 1000 kHz. The sensors which are mounted on the lower portion of the transformer tank have received signals belonging to cluster 1 and probable origin of these may be background noise. Sensors 2, 5, 8 and 16 have received signals of cluster 2. Sensors 2, 6, 8 and 16 mounted at the top side of the tank have received the signals belonging to cluster 0. Cluster 0 is isolated from other clusters for amplitude, energy and frequency with higher values and has a narrow frequency range as 55 to 110 kHz. Cluster 0, therefore appears to represent the acoustic PD activity in the transformer insulation.

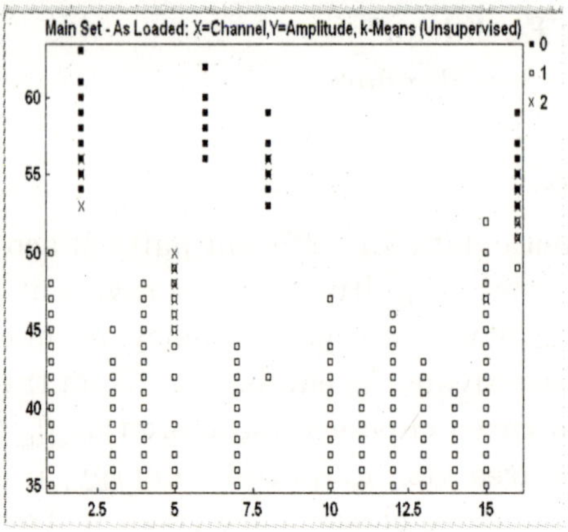


(a)

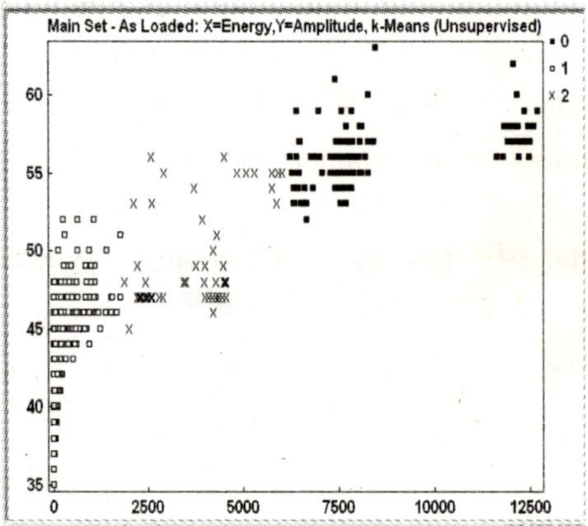


(b)

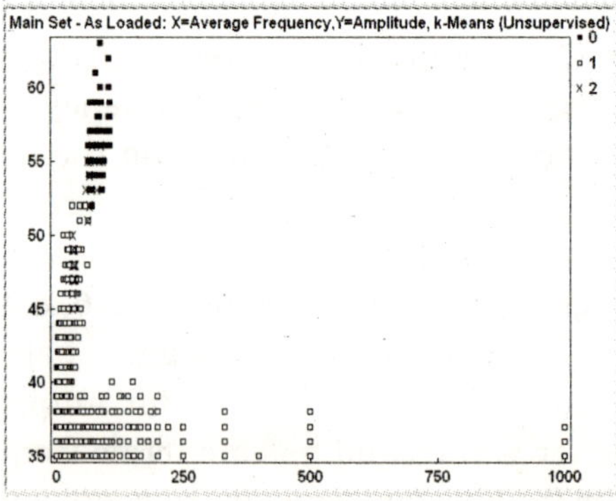


(c)

**Fig. 4.24** *(Contd...)*

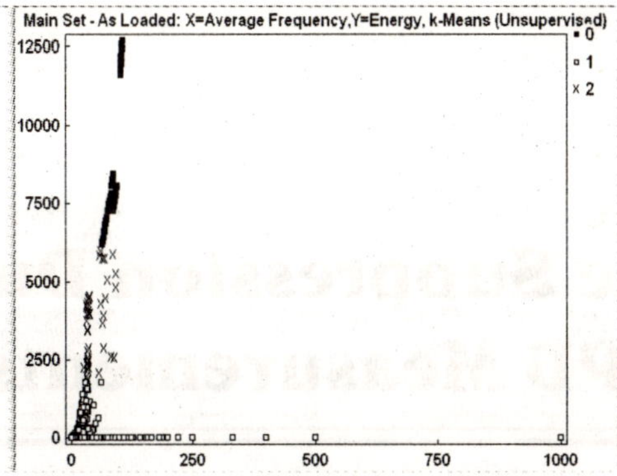


(d)

**Fig. 4.24.** Clusters for different acoustic features **(a)** Amplitude Vs channel, **(b)** Amplitude Vs Energy, **(c)** Amplitude Vs Avg. frequency and **(d)** Energy Vs Avg. frequency

CHAPTER

5

# Noise Suppression During PD Measurements

## 5.1 SOURCES OF INTERFERENCES IN PD MEASUREMENTS

The study of the origin, nature and manifestation of different sources of noise and their identification is an essential part of source description based on which a reasonable estimate of the amount of noise pollution can be made. Essentially, two broad types of interferences, which often pollute the PD signal, are;

i. Random internal noise

ii. Externally coupled disturbances

Random noise results from, for example, thermally stimulated currents and random fluctuations in electronic devices such as amplifiers, ICs, detection circuit impedance and the like. External disturbances, on the other hand, tend to be much more virulent than random noise. The external interferences, normally encountered in actual measurements particularly during site measurements can broadly be classified as follows:

i. Discrete Spectral Interferences (DSI): from radio transmission and power line carrier communication systems.

ii. Periodic Pulse-Shaped Interferences from power electronic circuits or during periodic switching of adjustable speed thyristor drives.

iii. Stochastic Pulse-shaped Interferences (SPI): from infrequent (aperiodic) switching operations, arcing between metallic contacts, arcing from slip-rings in rotating machines, Radio interference and corona from transmission lines and the like.

Some of the typical noise patterns encountered during PD measurements are depicted in Fig. 5.1. The block diagram in Fig. 5.2 summarizes the external interference and their sources.

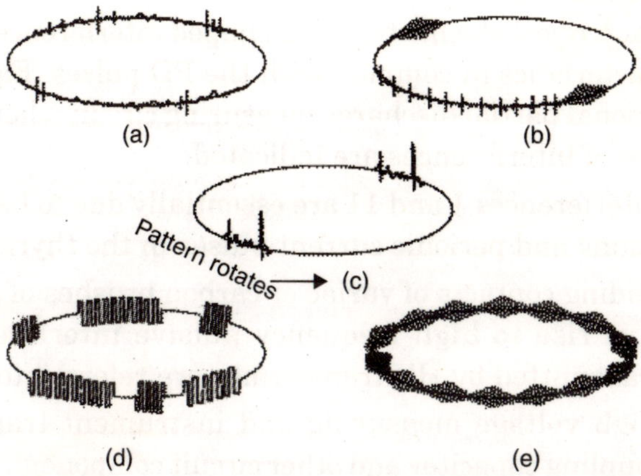


**Fig. 5.1.** Possible Types of Noise associated with PD caused by **(a)** Thyristor switching, **(b)** Fluorescent lamp, **(c)** Induction machine, **(d)** Time modulated (Morse) transmitter and **(e)** Broadcast transmitter or high frequency equipment

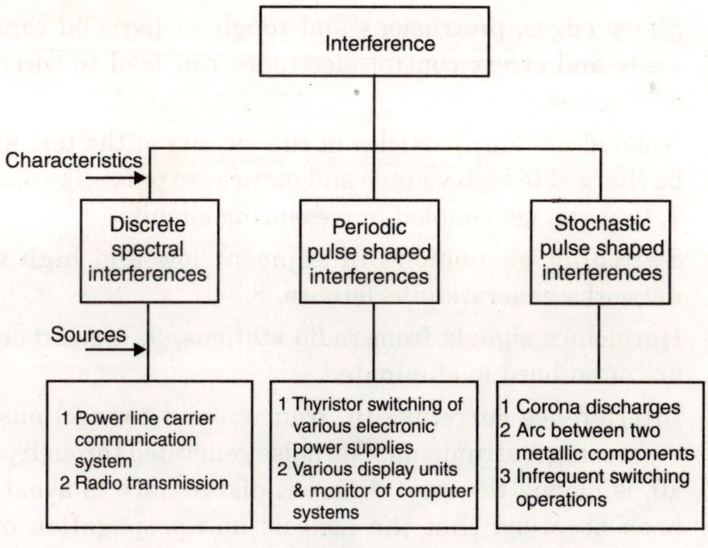


**Fig. 5.2.** Types of interference and their sources

Of the external interferences mentioned above, the continuous sinusoidal interference such as the DSI are easily recognized in the frequency domain, as they have narrow band frequency spectrum, concentrated around the dominant frequency and hence are simpler to eliminate. Periodic pulse shaped interferences can also be time-gated depending on their phase of occurrence. But, in effect, it is not so simple to identify and reject stochastic pulse-shaped interferences, as they have many characteristics in common with the PD pulses. Figure 5.3 shows the conventional partial discharge measuring circuit where the potential sites/sources of interferences are indicated.

i. Interferences 1 and 11 are essentially due to harmonics in the mains and periodic current pulses in the thyristors.

ii. Sliding contacts of variac or carbon brushes of generators can give rise to high frequency pulsive interferences that are transmitted by HV transformer are referred to as 2.

iii. High voltage measuring and instrument transformers, the coupling capacitor and other circuit components listed as 3 and 6, are also the possible sources of PD, corona and other parasitic disturbances.

iv. The interference on the high voltage side, 4, can effectively be blocked by a filter, provided no additional noise is generated by itself.

v. Sharp edges, protrusions and rough surfaces on connecting leads and stress control electrodes can lead to corona type discharges, 5.

vi. Isolated metallic particles in the vicinity of the test area can be charged to high voltage and may cause pulse-like discharge, 7, that can get coupled to measuring circuit.

vii. Switching phenomena in adjacent low and high voltage networks generate interference, 8.

viii. Harmonics signals from radio stations, 9, are airborne and are often hard to eliminate.

ix. Interference currents, 10, from ground connections of PD measuring instruments. The noise generated through position, 10, is among the most different disturbance to avoid. It has been observed that the genesis and propagation of these disturbances which are essentially low and medium frequency are unknown. Since the design of filters for frequencies encountered here are difficult, the mains borne disturbances are often a matter of great concern.

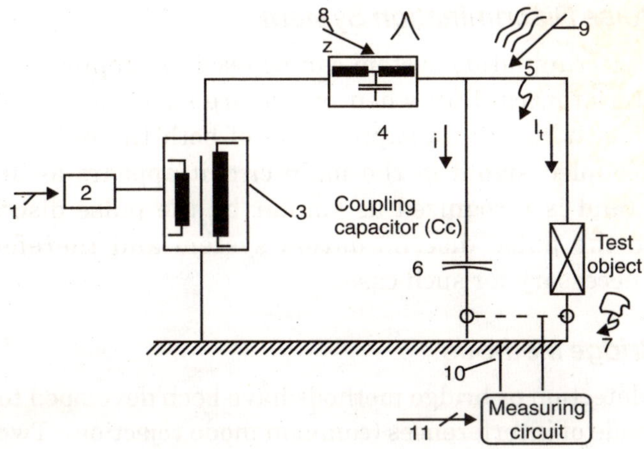


**Fig. 5.3.** Noise sources in a PD measuring circuit

## 5.2 INTERFERENCE SUPPRESSION METHODS

The simplest, but, not always a viable, way to eliminate the interferences, is to effectively shield the laboratory (Faraday Cage) and using appropriately designed filters for the electrical supply to the test circuit. But the use of filtering and screening techniques could be expensive or impractical, especially when testing large equipment. At the present, several methods of suppressing interferences have been adopted and these methods can broadly be classified into analog and digital methods.

### 5.2.1 Analog Methods

Analog noise filters, usually electrical models, are analogues of mechanical systems working in continuous time. An analog filter is a frequency selective filter, which operates on continuous-time signals. These are constructed using passive or active electrical elements, classified respectively as active and passive filters. Application of analog filters in extracting the PD signals is presented here under.

#### *5.2.1.1 Narrow Band Detection*

Narrow band detectors or "radio interference voltage measuring devices" (RIV meters as they are called) do suppress interferences to an extent, particularly those arising out from broadcasting stations and certain sinusoidal interference during on-site measurements. But the drawback of these methods is poor pulse resolution capability and lack of indication of the polarity of pulses.

#### 5.2.1.2 Pulse Discrimination System

The pulse discrimination system can be used to suppress the external pulsive interferences. But, when pulses are superimposed on pulsive interferences, it results in suppression of both the pulses. Also, any inductive coupled signal in the main circuit appears as an internal discharge (and is recognized as similar by the pulse discrimination system) to the pulse discrimination system and therefore makes screening necessary for such cases.

#### 5.2.1.3 Bridge Methods

Balanced detection or bridge methods have been developed to suppress common mode of interferences (common mode rejection). Two identical measuring impedances, used in adjacent arms of the bridge act in tandem to subtract, algebraically, the interferences appearing at the input of the amplifier, so that, a 20 dB improvement in DSI rejection can be achieved. In practice, however, a satisfactory bridge balance can only be achieved at a given frequency or within a narrow frequency range. A suggested modification to the bridge circuit, by inserting the measuring impedances at the HV end instead of the ground end, to prevent ground interferences mentioned earlier, seems to mitigate the problem to a certain extent. A noise canceller using delay line and use of tunable rejection filters to suppress DSI's suggested in the literature is found to be impractical. In the recent past, digital methods have found wider application, being relatively inexpensive, elegant and less cumbersome.

### 5.2.2 Digital Methods

With the advent of fast digital computers with large memory, digital methods for noise suppression in PD measurements, wherein, analogue data can be digitized and stored, are becoming more acceptable and preferable compared to the other methods. The important advantages of digital methods are;

- Possibility of post processing and reconstruction of PD data.
- Significant saving on the processing time.
- Feasibility of hardware implementation with possibility of real-time implementation.

Various digital signal processing techniques are now available for the suppression and even elimination of interferences in particular, as applied to PD measurements. The essence of some of these methods is presented hereunder.

#### 5.2.2.1 Filtering

A large body of literature is currently available wherein several methods of noise filtering have been described. Figure 5.4 lists the work covered by several authors in this area. Filtering is essentially a technique in time domain, noise denigration is affected in time domain itself. The advantage of the digital filters compared to the analog filters are, compactness, ease of design for desired frequency, multiple pass-band with sharp roll-off, adjustable order of configuration depending on the requirement and possibility of hardware implementation on a DSP processor.

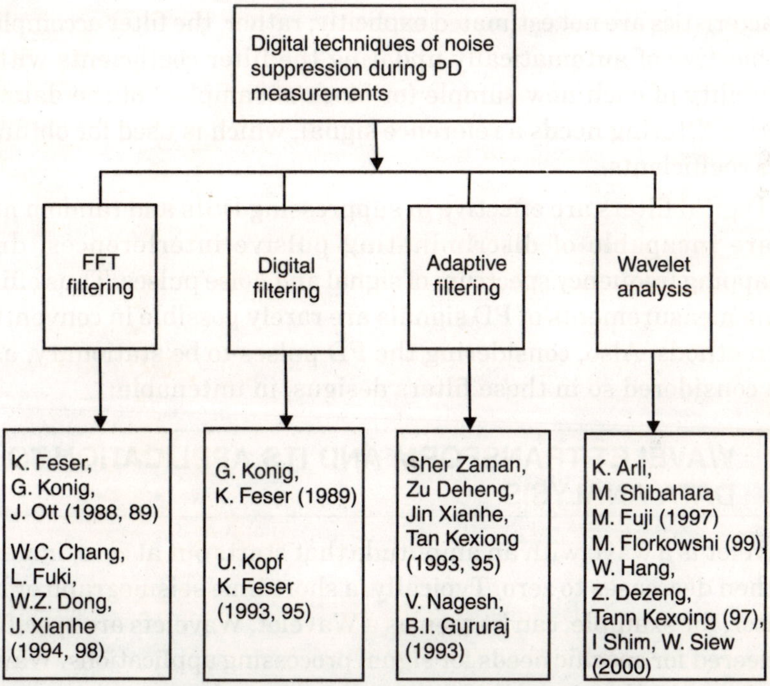


**Fig. 5.4.** Literature on digital techniques of interference suppression in PD measurements

In a broad sense, the methods of realization of digital filters can be divided into two classes as recursive filters and non-recursive filters. Recursive filters, also known as infinite impulse response (IIR) filters, used in realizing steeply falling filter characteristics. But, due to the regenerative feedback, the output can simply oscillate with resulting stability problem, particularly so with higher order filters. Further, larger phase distortions occur making reconstruction of signal in time domain difficult. Non-recursive or finite impulse response (FIR) filters are inherently stable and therefore do not result in oscillatory response. They have a fairly linear phase type and therefore distortion is relatively low.

Digital filters can be further classified as non-adaptive and adaptive filters. In non-adaptive methods, a small block of the recorded signal pack is initially used and its characteristics are calculated. Then, a fixed digital filter identity designed for eliminating the interferences optimally, shall be imposed on the block. In contrast, the adaptive methods have closed loop configuration, in which the filter coefficients are varied to reduce the interference. Here, the pertinent interference characteristics are not estimated explicitly; rather, the filter accomplishes the objective of automatically updating the filter coefficients with the availability of each new sample (or block of samples) of the data. The adaptive filtering needs a reference signal, which is used for obtaining, filters coefficients.

Digital filters are effective in suppressing DSIs and random noises but are incapable of discriminating pulsive interferences, due to overlapping frequency spectrum of signal and noise pulses. Thus efficient on-line measurements of PD signals are rarely possible in conventional DSP methods. Also, considering the PD pulses to be stationary, as are often considered so in these filters designs, in untenable.

## 5.3 WAVELET TRANSFORM AND ITS APPLICATION TO PD DATA ANALYSIS

A wavelet is a wave with an amplitude that starts out at zero, increases, and then decreases to zero. Typically, a short time seismograph or heart monitor, for example, can be seen as a Wavelet. Wavelets are specifically engineered for specific needs for signal processing applications. Wavelets can be combined, using a shift, multiplication and summation operation called convolution, to extract the information from the unknown signal. A wavelet is a mathematical tool which can be used to extract information from many different kinds of data, including - but certainly not limited to - audio signals and images. Sets of wavelets are generally needed to fully analyze data. A set of complementary wavelets can often be constructed and deconstructed, the deconstruction process being mathematically reversible. Thus, sets of complementary wavelets are useful in wavelet based compression/decompression algorithms where it is desirable to recover the original information with minimal loss.

Wavelet transforms are now being adopted for a number of applications, often replacing the conventional Fourier Transform. Important areas of applications are seismic geophysics, image processing, ECG signal processing and the like. An important use of wavelet approximation is in data compression. Like some other transforms, wavelet transforms can be used to transform data, then encode the transformed data, resulting in effective compression. A related use is that of smoothing/denoising data based on wavelet coefficient thresholding, also called wavelet shrinkage. By adaptively thresholding the wavelet coefficients that correspond to undesired frequency components, smoothing and/or denoising operations can be performed. Wavelet transforms are broadly divided into three classes: continuous, discrete and multi-resolution-based. In continuous wavelet transforms, a given signal of finite energy is projected on a continuous family of frequency bands (or similar subspaces, $L^2(R)$). For instance the signal may be represented on every frequency band of the form $[f, 2f]$ for all positive frequencies $f > 0$. Then, the original signal can be reconstructed by a suitable integration over all the resulting frequency components. The frequency bands or subspaces (sub-bands) are scaled versions of a subspace at scale 1. This subspace in turn is in most situations generated by the shifts of one generating function, $\psi \, \varepsilon \, L^2(R)$, the mother wavelet [75]. The projection of a function $x$ onto the subspace of scale $a$ has the form

$$x_a(t) = \int_R WT_\psi \{x\}(a,b).\psi_{a,b}(t)db \tag{5.1}$$

with wavelet coefficients

$$WT_\psi \{x\}(a,b) = (x, \psi_{a,b}) = \int_R x(t)\, \overline{\psi_{a,b}(t)}\, dt \tag{5.2}$$

It is computationally impossible to analyze a signal using all wavelet coefficients, so one may wonder if it is sufficient to pick a discrete subset of the upper half-plane to be able to reconstruct a signal from the corresponding wavelet coefficients. One such system is the affine system for some real parameters $a>1$, $b>0$. The corresponding discrete subset of the half-plane consists of all the points $(a^m, n\, a^m b)$ with integers $m, n\ \varepsilon\ z^2$. The corresponding *baby wavelets* are now given as

$$\psi_{m,n}(t) = a^{\frac{-m}{2}} \psi\left(a^{-m}t - nb\right) \tag{5.3}$$

A sufficient condition for the reconstruction of any signal $x(t)$ of finite energy by the formula

$$x(t) = \sum_{m\varepsilon Z} \sum_{n\varepsilon Z} (x, \psi_{m,n}).\psi_{m,n}(t) \tag{5.4}$$

is that the functions ($\psi_{m,n}$: $m,n \in Z^2$) and form $L^2$ (R).

For practical applications, and for efficiency reasons, one prefers continuously differentiable functions with compact support as mother (prototype) wavelet (functions). However, to satisfy analytical requirements (in the continuous wavelet transform) and in general for theoretical reasons, one chooses the wavelet functions from a subspace of the space $L^1(R)\ L^2(R)$. This is the space over which Eqns. 5.1 and 5.2 are absolutely square integrable:

$$\int_{-\infty}^{\infty} |\psi(t)| dt < \infty \text{ and } \int_{-\infty}^{\infty} |\psi(t)|^2 dt < \infty$$

Being in this space ensures that one can formulate the conditions of zero mean and square norm one:

$$\int_{-\infty}^{\infty} \psi(t) dt = 0$$

is the condition for zero mean, and

$$\int_{-\infty}^{\infty} |\psi(t)|^2 dt = 1$$

is the condition for square norm one.

For $\psi$ to be a wavelet for the continuous wavelet transform, the mother wavelet must satisfy an admissibility criterion (loosely speaking, a differentiability criterion) in order to get a stable, invertible transform.

The wavelet is entirely defined by the scaling filter - a low-pass Finite Impulse Response (FIR) filter of length $2N$ and sum 1. In bi-orthogonal wavelets, separate decomposition and reconstruction filters are defined. For analysis the high pass filter is calculated as the quadrature mirror filter of the low pass, and reconstruction filters the time reverse of the decomposition.

Wavelets are also defined by the wavelet function $\psi(t)$ i.e. the mother wavelet and scaling function $\phi(t)$, the father wavelet, in the time domain. The wavelet function is in effect a band-pass filter and scaling it for each level halves its bandwidth. This creates the problem that in order to cover the entire spectrum, an infinite number of levels would be required. The scaling function filters the lowest level of the transform and ensures the entire spectrum is covered. For a wavelet with compact support, $\phi(t)$ can be considered finite in length and is equivalent to the scaling filter $g$. The wavelet only has a time domain representation as the wavelet function $\psi(t)$.

### 5.3.1 Method of Wavelet Transform

This technique analyzes the signal both in time and frequency domain simultaneously. A wavelet transform is a linear transform, which can treat non-stationary signals. Just as Fourier transform decomposes a signal into sinusoidal waves of various frequencies, wavelet transform breaks up a signal into shifted and dilated versions of what is called a mother wavelet. The fixed time frequency resolution of short-time Fourier transform (STFT), which can be thought of as a possible method of analysis, has certain constraints in regard to analysis of non-stationary signals. Also, significant possibility of generating fast and computationally better algorithms of Decimated Wavelet Transform (DWT) makes it more suitable for the present purpose than STFT. The mother wavelet is a window function, whose extent enlarges in time (or reduction in frequency), while resolving low-frequency component and shrinks while analyzing high frequency components.

The method of denoising of a polluted signal involves two steps. In the first phase a desired mother wavelet is chosen according to the characteristics of the signal and then DWT coefficients are obtained. The second phase involves the modification of these wavelet coefficients using hard or soft thresholding. The denoised signal is then reconstructed by taking inverse wavelet transforms of modified components. Figure 5.5 illustrates a denoising process by this method.

## 5.4 MOVING AVERAGE (MA) TECHNIQUES

Measured data pertaining to a physical process most often contain other components which are not generic to data. In a very broad sense, these components can be called as "Noise" as opposed to the actual or the expected data which may be called as the signal. Filtering is a process by which unwanted or noise or disturbances or interference in a real time data, are either suppressed or if possible eliminated. Measurements on time dependent process post an array of outputs varying in magnitude. We may call this as a Time Series (TS) data set, shown in Fig. 5.6.

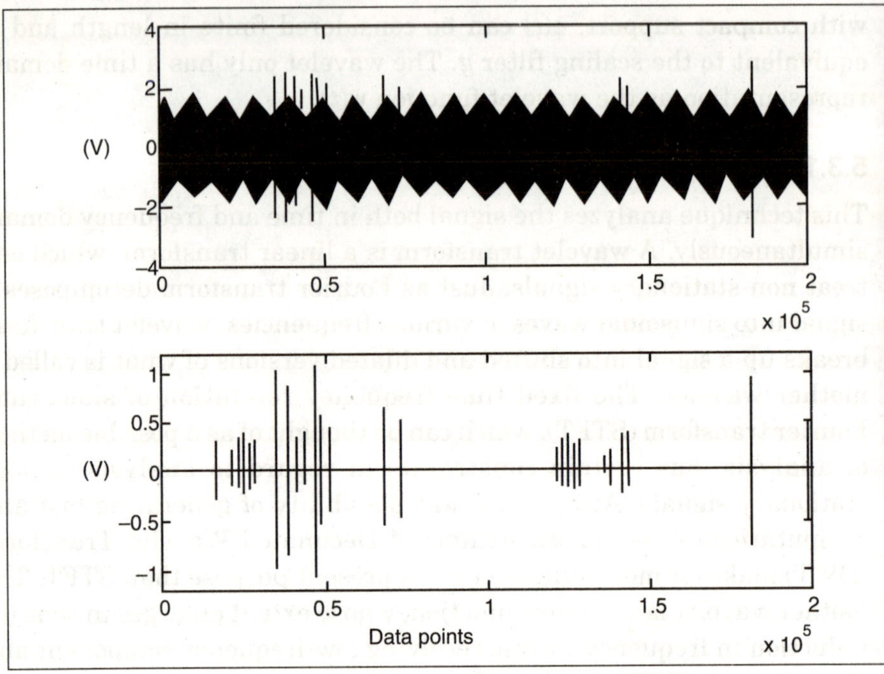


**Fig. 5.5.** Enhancement of simulated signal using DWT approach: (i) Simulated noisy PD signal and (ii) Enhanced signal

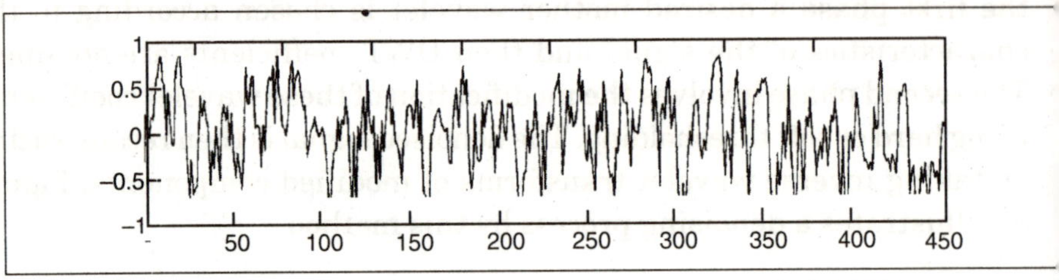


**Fig. 5.6.** Time series data on sunspot activity measured on a monthly basis [76]

If $x(t)$ is the data realized at any instant, $t$, of time, this always contains the noise $\varepsilon_t$. If $u(t)$ is the actual measured data, then we may formalize this statement as

$$u(t) = x(t) + \varepsilon_t \tag{5.5}$$

The term $\varepsilon_t$ has different connotations when applied to different situations, for example, noise, error and so on. Also, if $\varepsilon_t = c$, a constant, then, in some way, the noise can be estimated and the signal can be retrieved easily. The Eqn. (5.5) is referred to as a time series. Paired series of data $\{t_i, u(t_i)\}$ is often very large in which, $x(t)$ may be a continuous component or may be present only over a limited interval of time.

It is also possible that the data may be discontinuous, such as, for example, characterized by gaps where no measurements have made or time windows over which the data has not been reported as shown in Fig. 5.7.

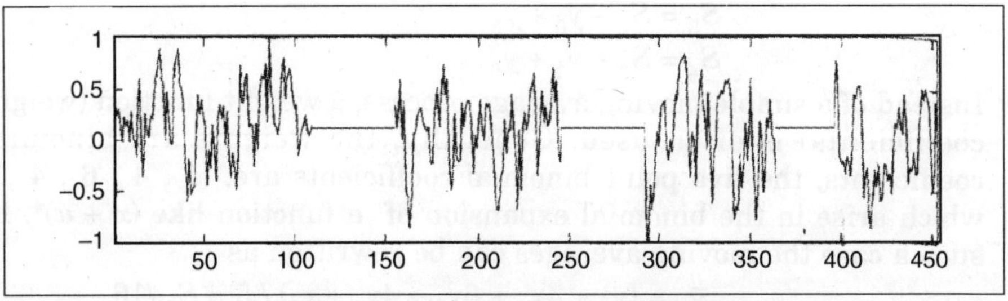


**Fig. 5.7.** Time series containing gaps

Often need arises of filling the gaps with data which, we have reason to believe, "belongs to" the existing data pockets.

Terms such as "trend", "error", "signal- to- noise ratio, S/N" & "moving average" are encountered in connection with time series analysis. Moving average refers to the process of smoothing the time series data (TSD). This involves time-windowing the TSD, the window may either be of equal or unequal lengths. If the data in each window is averaged and this is added on to the data in the next window & so on, a much smother pattern is obtained. Also if the noise accompanying the data is bi-directional & perfectly random, then, the process of averaging entails a cancellation of noise thereby preserving the signal. This means, statistically speaking, the noise has a zero-mean value & a characteristic variance [77] which is nearly invariant with respect to time. This is the moving average process.

The moving average models are applied to data processing in various situations. Two possible applications are described herein to indicate its usefulness. The first illustration deals with a analytical fitting exercise wherein a time series dataset is subjected to a moving average procedure with a view to extrapolate the data to later periods of time. Suppose, for convenience, successive observations on a random variable, $y$, are assumed to have been made at equal intervals of time. The time, an independent variable, is often called as a hidden variable. Let the successive values of the process under considerations be $y_0$, $y_1$ etc. A simple moving average process taking, for example, first five successive values of $y$, can be written as;

$$\bar{y}_2 = (y_0 + y_1 + y_2 + y_3 + y_4) / 5 = S_2 /5$$

$$\bar{y}_3 = (y_1 + y_2 + y_3 + y_4 + y_5) / 5 = S_3 /5$$

$$\bar{y}_4 = (y_2 + y_3 + y_4 + y_5 + y_6) / 5 = S_4 /5$$

These equations suggest, that, $y_2$ ..... $y_4$, are the mean values of the five points identified in the equations and are successively placed at the time delays $t = 2$, $t = 3$, and so on. In practice, it is usually quicker to rewrite the successive sums for example as:

$$S_3 = S_2 - y_0 + y_5$$
$$S_4 = S_3 - y_1 + y_6$$

Instead of a simple moving average process, a weight function (weight coefficients) is often used. Generally, the weights are binomial coefficients, the five point binomial coefficients are: {1 , 4 , 6 , 4 ,1} which arise in the binomial expansion of a function like $(x + a)^4$. In such a case the moving averages can be rewritten as

$$\bar{y}_2 = (y_0 + 4y_1 + 6y_2 + 4y_3 + y_4) / 5 = S_2 /16$$
$$\bar{y}_3 = (y_1 + 4y_2 + 6y_3 + 4y_4 + y_5) / 5 = S_3 /16$$
$$\bar{y}_4 = (y_2 + 4y_3 + 6y_4 + 4y_5 + y_6) / 5 = S_4 /16$$

The weighted coefficients have a tendency to produce a better smoothing procedure while preserving the main feature of the main series. One of the disadvantages of moving average technique is that, a $(2k + 1)$ time series entails the loss of the $k$ terms at the beginning and end of the series. However, when the time series is very long this loss is largely made up by the remaining terms. The following data-set is given to illustrate the computational procedure.

In Tables 5.1 and 5.2, the dissipation factor of a 18kV 100 MVA rotating machine stator insulation and PD data acquired on a RIP transformer bushing over a month's duration are included as examples. The machine insulation was F – Class (maximum operating temperature = 155°C). The measurements have been performed under normal operating conditions and at intervals of one day, spread over a thirty one day cycle. In Figs. 5.8 and 5.9, the raw data of tan δ and PD, the moving average (computed on a five point mean basis) and a 'Notional Trend Line' fitted to the moving average are shown. The trend line is a second degree polynomial and is seen to fit the data to a reasonable degree of accuracy.

**Table 5.1.** Dissipation factor, tan δ ($10^{-3}$) as a function of time

| Days | 1 | 2 | 3 | 4 | 5 | 6 | 7 | 8 | 9 | 10 |
|---|---|---|---|---|---|---|---|---|---|---|
| tan δ ($10^{-3}$) | 89 | 87 | 81 | 91 | 96 | 99 | 98 | 112 | 118 | 123 |

| Days | 11 | 12 | 13 | 14 | 15 | 16 | 17 | 18 | 19 | 20 | 21 |
|---|---|---|---|---|---|---|---|---|---|---|---|
| tan δ ($10^{-3}$) | 96 | 122 | 158 | 162 | 162 | 166 | 156 | 142 | 173 | 164 | 180 |

| Days | 22 | 23 | 24 | 25 | 26 | 27 | 28 | 29 | 30 | 31 |
|---|---|---|---|---|---|---|---|---|---|---|
| tan δ ($10^{-3}$) | 156 | 136 | 143 | 213 | 217 | 181 | 166 | 162 | 167 | 213 |

**Table 5.2.** PD as a function of time

| Days | 1 | 2 | 3 | 4 | 5 | 6 | 7 | 8 | 9 | 10 |
|---|---|---|---|---|---|---|---|---|---|---|
| PD, pC | 570 | 580 | 640 | 710 | 740 | 690 | 820 | 860 | 700 | 900 |

| Days | 11 | 12 | 13 | 14 | 15 | 16 | 17 | 18 | 19 | 20 | 21 |
|---|---|---|---|---|---|---|---|---|---|---|---|
| PD, pC | 940 | 860 | 610 | 610 | 590 | 820 | 850 | 880 | 940 | 990 | 790 |

| Days | 22 | 22 | 23 | 24 | 25 | 26 | 27 | 28 | 29 | 30 |
|---|---|---|---|---|---|---|---|---|---|---|
| PD, pC | 890 | 890 | 950 | 900 | 890 | 970 | 990 | 1060 | 980 | 1140 |

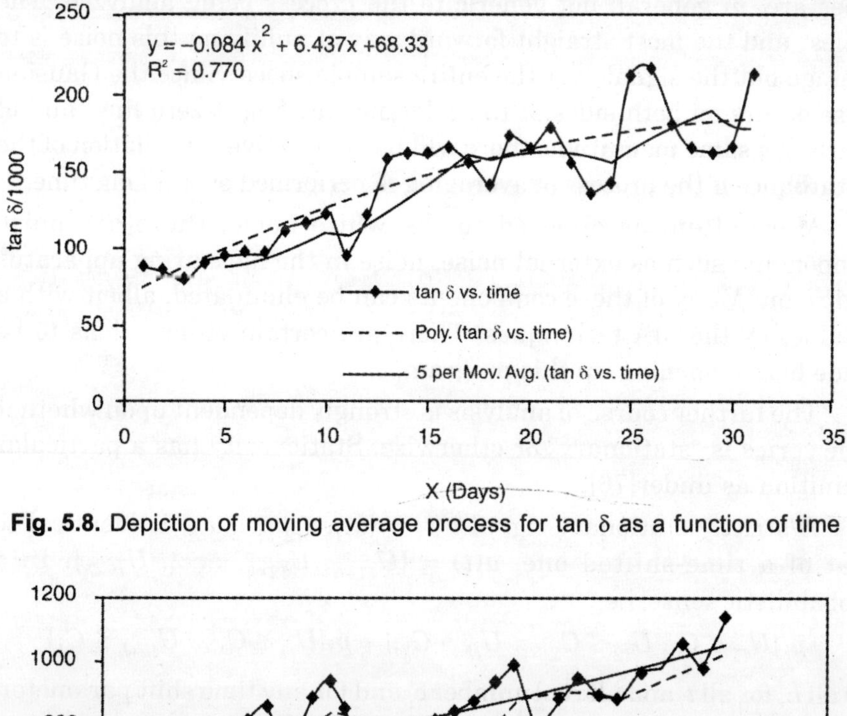


**Fig. 5.8.** Depiction of moving average process for tan δ as a function of time

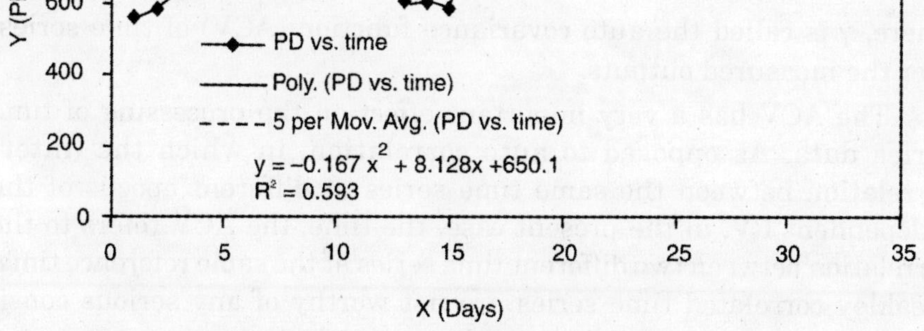


**Fig. 5.9.** Depiction of moving average process for PD as a function of time

It is important to note that a polynomial fit of any degree to the raw data carries with it a considerable degree of variability and hence a large deviation in the expected values of the parameter estimated. Concomitantly, any estimate made on the probable life of the equipment in question becomes correspondingly less accurate.

As a second application, a time series is analyzed for estimating and eliminating the systemic noise. There are many important concepts involved in noise suppression by this method. Among others is the consideration of the white noise. The Gausian occurrence of certain noise types are, in general, not generic to the process being analyzed. The easiest and the most straight forward way of nullifying this noise is to average out the signal over the entire sample space. Since the Gausian noise occurs on both sides of the reference (calling it zero line) and of nearly the same magnitude, there will be an effective cancellation of the disturbance if the process of averaging is performed over a long time.

More often, as opposed to the white noise, there are noise components such as external noise, noise in the measuring apparatus and so on. Many of these components can be eliminated, albeit with a residue, by the MA techniques. There are certain observations to be made before•opening up this program.

The further course of analysis is strongly dependent upon whether time series is "stationary" or otherwise. Stationarity has a particular definition as under[76],

A strictly stationary signal $u(t) = \{u_{t1}, u_{t2}.......u_{tk}\}$ is identical to that of a time-shifted one, $u(t) = \{U_{t1+h}, U_{t2+h}, \dots \dots U_{tk+h}\}$. In a probabilistic sense, i.e.:

$$p\ \{U_{t1} \leq C_1, U_{t2} \leq C_2, \dots U_{tk} \leq C_{k}\} \equiv p\ \{U_{t1} \leq C, \dots U_{tk+h} \leq C_k\}$$

for all $k$, for all $t_i$ and all real number $c_i$ and for any time shift parameter, $h$. Suppose the data is windowed and consider $u_t$ and $u_{t+h}$

$$\gamma\,(s,t) = r\,(s + h, t + h)$$

where, $\gamma$ is called the auto covariance function (ACV) of time series, over the measured outputs.

The ACV has a very important effect on the processing of time series data. As opposed to auto correlation, in which the (inter) correlation between the same time series at different epochs of the independent RV, in the present case, the time, the ACV refers to the correlation between two different time series at the same reference time. Weakley correlated time series are not worthy of any serious consideration in noise suppression paradigms.

In insulation statistics, a time and stress dependent property is carefully examined for theses statistical parameters so as to be able to check whether a series of measurements, at constant stress, depicting temporal changes is homogeneous or not. If the homogeneity cannot be guaranteed, then, the parameter estimates are in doubt, or the confidence intervals of the estimated parameters would become so large as to make a meaningful assessment on the status of insulation frivolous. The expectation operator, $E$, permits writing the auto covariance function as:

$$\gamma(t + h, t) = E\,[(U_{t+h} - \mu)\,(U_t - \mu)]$$

where, $\mu$ is the mean of the stationary time series (STS). Observe that it is independent of $t$. Using this fact, the auto covariance of a stationary time series is written as:

$$\gamma(h) = E\,[(U_{t+h} - \mu)\,(U_t - \mu)]$$

With similar notations, the variance of a time series at $t$ and at $t+h$ can easily be visualized.

The auto-correlation function,(ACF), $\rho(h)$ has the same meaning as the well known correlation function and is written as:

$$\rho(h) = \frac{\gamma(t+h,t)}{\left[\sqrt{\gamma(t+h,t+h)\gamma(t,t)}\right]}$$

The cross covariance function (CCF) of a pair of jointly stationary (each being stationary) time series is written as:

$$\gamma_{u,v}(h) = E\,[(U_{t+h} - \mu_u)\,(v_{t+h} - \mu_v)]$$

The data processing involves the following steps:

i. Obtain the analogue signal.

ii. Input (i) to computer and digitize.

iii. Program the expression for ACV, ACF and CCF.

iv. Obtain the M.A. spectrum ( obviously there will be a loss of signal component and introduction of a parasitic noise component).

The moving average spectrum gives the most probable signal component bereft of much of the noise. The basic principle used in moving average method is to affect a time average of a certain portion of the signal sequence at a known instant of time. This essentially is a synchronous accumulation of the quasi-repeatable signals using the information of either repeated polarity, or a characteristic of random noise, hence effectively cancelled it in time. More explicitly, periodic signals are added together coherently, while the random noise components are averaged out to successively diminishing values. The improvement in signal–to–noise ratio (SNR) by this method is $\sqrt{N}$, where $N$ is the number of samples to be arranged. The term coherent here has the same meaning as optical coherence.

Moving average method is simple, easy to implement and effective in removing white noise. The efficiency of this method is reduced considerably in the presence of multiple DSIs. This method is not useful in on-site measurements and non-shielded HV labs, since the signal to noise ration of such signals are extremely low and PD pulses buried in such high noise is non-tractable.

Arguably, one of the more important disadvantages of the moving average model is that the degree of smoothing of the variant data may seem unacceptably high and there may be situations in which a better part of the signal due actually to the physical process is substantially played down. Also, depending on the length of the time series, the loss of first few measurements during the process of averaging may, some times, prove to be detrimental to the estimates. The method of moving average has been demonstrated in Fig. 5.10 taking an example of actual surface temperature of an ocean (Fig. 5.10 (a)) and its 7-point moving average.

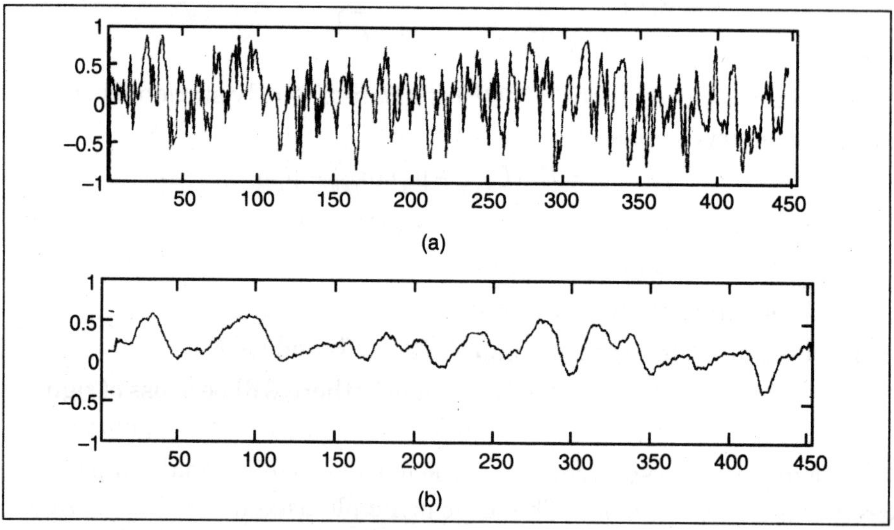


**Fig. 5.10.** A typical Ocean surface temperature data recorded on a monthly basis **(a)** Actual data and **(b)** Moving average profile of (a) [76]

CHAPTER

6

# Equipment Specific Diagnosis and Reliability Assessment

## 6.1 GENERAL

The general procedure for the assessment of the condition of power apparatus insulation at any instant of time involves the enunciation of a failure criterion described later. The principal stress causing degradation of insulation is being different for different equipment, the strategy for Diagnostic Testing and Condition Monitoring (DTCM) specific to any particular equipment is to be worked out keeping in view this important aspect. Furthermore the class, kind and quality of insulating material used in equipment is also a matter of consideration. In this chapter, the diagnostic test methods to be used in each of such cases are detailed.

## 6.2 TYPES OF INSULATION SYSTEMS USED IN POWER EQUIPMENT

According to the type of insulation used, which, in turn is dictated by the environment in which the equipment is situated, two broad variety thereof can be mentioned. They are, Dry Type Insulation (DTI) and the Liquid Impregnated Systems (LIS). However, gas impregnated solid insulation is covered under the head, DTI. The more important high voltage apparatus using dry type insulation are the following;

i. Extruded HDPE and XLPE insulated Cables for Low, Medium and High Voltages.

ii. Epoxy Resin Bonded, mica or glass fibre-based insulation for High Voltage Rotating Machines, in particular, Generator Stators.

iii. Liquid-free Power Capacitors.

iv. Support insulators in Gas Insulated Switchgear.
v. Synthetic Resin Bonded Paper Bushings for Power Transformers, PT's, CT's and Wall Bushings.
vi. Epoxy Insulated Foil Wound Transformers for Low and Medium Voltages.
vii. Outdoor Epoxy-cast insulators and supports.
viii. Synthetic and Natural Rubber composites.

The processes leading to long time ageing failure of DTI with or without gas filling are considerably different from those of LIS. Also, since invasive tests of any kind are entirely precluded, the diagnosis in such situations is made based on external terminal electrical measurements and/or through enclosure mounted non-electrical sensors. The results of such tests indicate a possible defect, but, the location is not easy to ascertain. The more important tests prescribed in the DTCM of DTI are the following:

i. Visual inspection, if the insulation is exposed, like for example, outdoor line insulators and supports.
ii. Dissipation factor Tip-up measurements.
iii. More importantly, PD measurements, both electrical and acoustic emission methods.
iv. Ultrasonic tests.
v. Infrared thermal imaging for mapping the spatial and temporal temperature distribution.

## 6.3 EQUIPMENT SPECIFIC DIAGNOSTIC TESTING AND CONDITION MONITORING

Condition monitoring of apparatus insulation during service through relevant non-destructive test methods is a subject of great importance. As has already been pointed out, diagnostic tests reveal the status of insulation at some given instant of time besides offering a possible means of estimating, approximately, the remaining useful life of power apparatus. Depending upon the type and nature of the insulation system, different diagnostic procedures are prescribed to identify the parameters indicating the status of insulation. Any developing defect or an abnormality in the insulation behaviour can be diagnosed early enough by properly structuring the frequency of tests. In the beginning when the equipment is put into service, frequent tests are required to weed out the infantile irregularities. This regime normally covers the first six months to one year after the equipment has been put to service. Planning the inspection time window is a statistical exercise and should be undertaken with the prior information.

### 6.3.1 Inspection Time Slot

In deciding the inspection time window, it is important to know if the insulation undergoes a destructive or a non-destructive test. When only diagnostic tests are performed, the time windows could be narrowed so that, over the life time of a particular equipment under test, the changes in the diagnostic property and hence the ageing behaviour could be assessed and a large amount of data can be gathered. On the other hand as in the laboratory experiments if the specimens are run to failure (destructive testing) each time they are tested, we would have only one data point for that specimen, and assuming that identical specimens are used, to collect fairly accurate data a large number of specimens are required. Interruption of experimental run is another consideration. During testing, it is assumed that the insulation is not aged. But if the testing time is larger, there is a finite possibility of improvement or degradation in some of the diagnostic properties. If a specimen is found to have failed at a certain time, it is customary to reckon the time to failure as the time up to the previous inspection slot. When the time windows are too wide, the data point obtained in this way would estimate the life pessimistically.

Again, inspection time window has very serious implications on the cost and time factors in ageing experiments. Therefore, the decision on the inspection time slot should be taken only after running a series of preliminary ageing experiments, which also help in the fixation of appropriate ageing stresses.

The term 'failure' in connection with ageing experiments should be carefully defined. A complete electrical short circuit of insulation through its body because of which, it cannot support the rated applied electrical stress is actually defined as failure. For this condition to reach, if it reaches at all, an ageing experiment has to be run over inordinately long periods of time thus defeating the purpose of running economical ageing experiments. When insulation ageing studies are made under stresses other than electrical, like thermal, mechanical and so on, the above definition of failure cannot be applied. One then considers a property other than electrical, say loss of mechanical strength, loss of mass, and the like and failure is defined as complete loss of that property.

Even here, the constraint on experimental time should be kept in mind. To mitigate this situation, the concept of subjective failure is applied. This means that there is no need to wait until a certain property, electrical or otherwise acquires an unacceptable value, the experiment will be terminated when a particular property reaches a specified lower limit say 30% or 50% of its original level. This means that the specimen is deemed to have failed even though it still possesses about 50% of the property. If this procedure is applied to all specimens, it is possible to obtain a comparative idea of the life. Fixation of this lower limiting value of a property to define failure is a difficult proposition calling for sound engineering judgment and also has overtones of economic implications.

To uniformize the failure criteria and to quantify the degradation, the following procedure is adopted.

i. In the series of experiments under electrical or mechanical or thermal stress, a certain number of specimens put to ageing are drawn at random at any inspection time window and their dielectric strengths are measured. If the scatter in the measurement is small, a mean value of the measurement determines the dielectric strength. When this drops to 50% of the initial value, the corresponding time slot is taken as the time to failure of that sample size.

ii. When the stresses used are mechanical or thermal in nature, either consider electrical property or mechanical strength and define failure in the same way as above.

Before planning diagnostic tests, it is necessary to identify and understand the possible modes of failure of insulation in question during laboratory tests and under normal operation conditions. In practice, it may happen that, the correlation between these two sets of data is poor leading to erroneous conclusions being drawn on the remaining useful life of power apparatus. For example, there may be preferred areas or zones of failure in the body of the equipment, where failure might occur as a result of excessive stress concentration. Also, in the same equipment, failure can ensue in different components and at different locations depending upon the form of stress, alternating, direct or impulse. A prior knowledge of these aspects would greatly enhance the usefulness of a DTCM programme.

Essentially two types of insulation systems are encountered in practice; the dry type and the insulating liquid impregnated type.

## 6.4 DRY TYPE SYSTEMS

Dry type systems are solid insulation systems in which liquid phases like an impregnant or internal coolant is not present. However, gas impregnated solid insulation is covered under this head. The more important high voltage apparatus using the insulation under this head are:

i. Extruded Polymeric Cables (XLPE)
ii. Dry-Type Transformers
iii. High Voltage Rotating Machine Stator Insulation
iv. Gas Insulated Substations

### 6.4.1 Power Cable Insulation

Low and medium voltage power and distribution cables form a very important component of electrical energy transmission and distribution system. Starting from the time the underground cables are laid, a series of tests are performed on the cable itself and on other components like, joints, terminations and connectors to overhead lines called potheads.

#### *6.4.1.1 Modes of Failure of Synthetic Dielectric Cables*

During and immediately after manufacture, the power cables are subjected to a series of quality assurance tests at the works. Details of these tests are covered later in this section. After satisfactory results of proof tests against manufacturing defects, the cables and accessories are carried to the site. During transport and laying of the cables, internal damages may be caused to the insulation like, for example, strands of power conductor or metal sheath may bite into the dielectric forming a conducting defect, Fig. 2.1 (*a*). Also, in the manufacture, polyethylene cables are sometimes subjected to a steam curing cycle to relieve internal strains and annealing. During this process, minute, water filled and / gas filled cavities are formed, Fig. 2.1(b). Normally, in a well-manufactured cable, the physical size and the number of such inclusions per meter length of the cable are small. Failure of cable systems at their terminations, or at extension joints is also included under this head.

Diagnostic test on cables are usually conducted on complete installed cable circuits and include accessories (joints and terminations). Testing and assessment of installed cables are often undertaken for reasons other than condition monitoring, for instance for conducting acceptance test after laying and joint repairs or for effecting mandatory diversions.

For testing cables in-situ, several problems are encountered for example, the length of the cable, involvement of joints of various types, transportation of the measuring equipment to the site and requirement of large power if the cable length is very long. Another important aspect is that of ambient electro-magnetic interference, a factor needs special consideration in conducting such tests as partial discharges.

### 6.4.2 Recommended Diagnostic Tests for Extruded Power Cables

Among the more important functional tests found to be essential in DTCM are:

i. DC withstand/proof tests
ii. Polarization Index/Insulation Resistance Measurement
iii. Oscillating Wave Tests
iv. AC testing at
    a. Power Frequency
    b. Very Low Frequency (between 0.01 and 10 Hz)
    c. Partial Discharge Measurements at these frequencies
    d. Dissipation Factor Measurement

#### *6.4.2.1 DC Withstand/Proof Tests*

DC testing has long been used to ascertain the safety of circuits in a distribution network. The power source used is a static High Voltage (5-50 kV) DC generator capable of delivering charging currents of about 1 A. Although the information gained by testing with DC is only qualitative, nevertheless gross faults to ground are easily detected by DC proof testing up to the required voltage levels.

A caution must be exercised during this testing to see that the cable system is completely discharged. Otherwise, electric field due to the remnant charges adds to the alternating field upon subsequent energization. Also there is some fear that the repeated DC tests may endanger the cable system and reduce the service life. DC testing methods also include sheath tests on cables to ensure the integrity of the outer sheath. Any areas of high leakage can be located and repaired which reduces the chance of corrosion and failure.

#### 6.4.2.2 Polarization Index/Insulation Resistance Measurement

Severe internal erosion and moisture absorption/inclusion in the insulation body and erosion of cable grading material can be detected by measurement of PI and IR. Although no definite values can be quoted, IR and PI of respectively about 1000 MΩ and between 1.5 and 2.0 are easily realized. A point of particular interest in this test is that, if there are measurable changes in cable insulation during the DC test, they are readily shown up by a lowering of PI and IR values. Recovery voltage measurement serves as a broad guideline for charge leakage in insulation which again can be translated into an equivalent PI measurement [78].

#### 6.4.2.3 Oscillating Wave Test

An alternative to a standard DC test is the use of oscillating waves. This method has been shown to be successful in the detection of defects in polymeric cables, producing similar breakdown levels as mains frequency testing. The advantage here is that, since the wave form is oscillatory and persists for a very short duration, it produces similar effects as AC and the insulation is not stressed for a long duration unlike in the case of DC test. The damage done to the cable insulation system, if any, is minimal. A schematic diagram of Oscillating Wave Tests (OWT) [79] is shown in Fig. 6.1.

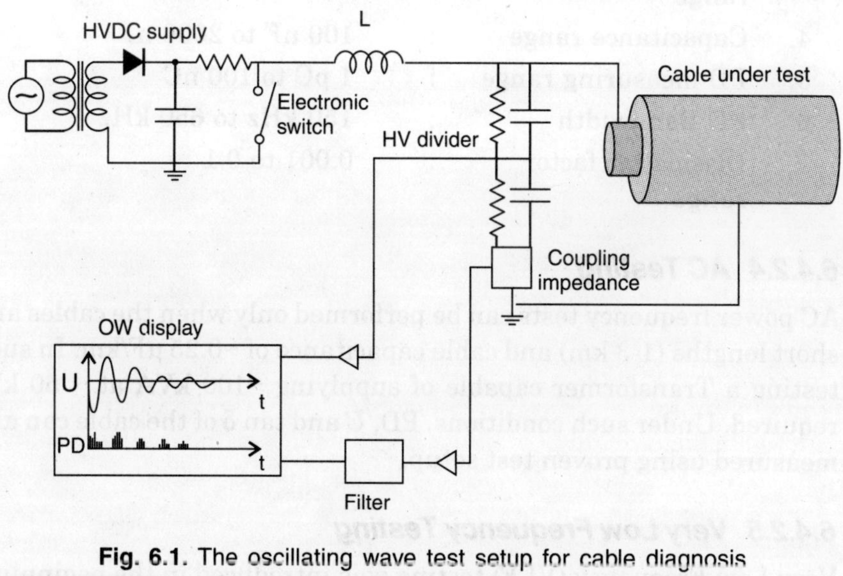


**Fig. 6.1.** The oscillating wave test setup for cable diagnosis

The OWT set generates and applies a damped oscillatory (decay time of 0.3 to 1 second) wave of frequency, *f* given by,

$$f = \frac{1}{2\pi\sqrt{LC}} \tag{6.1}$$

where $C$ is the capacitance of the test object (Cable) and $L$ is the inductance of an air cored tuning inductor. The value of $L$ is so selected as to obtain the frequency of oscillation between 50 Hz and 1 kHz. The $Q$ of the circuit is generally in the range of 30 and 100.

The PD is ignited in this frequency range and hence amply covers the power frequency. This aspect is in compliance with IEC 60270 recommendations. Location of PD pulse can be identified treating the cable as a transmission line and using the travelling wave techniques. Values of capacitance $C$ and tan $\delta$ can be calculated based upon the oscillating wave time and frequency. The typical values of the components of the PD circuit are provided here for reference.

1. Maximum voltage : 40 kV DC/ 30 kV rms
2. Coil inductance and resistance : ~ 0.7 H and < 20 Ω
3. Operating frequency range : 50 Hz to 1 kHz
4. Capacitance range : 100 nF to 2000 nF
5. PD measuring range : 1 pC to 100 nC
6. PD Bandwidth : 150 kHz to 650 kHz
7. Dissipation factor range : 0.001 to 0.1

#### 6.4.2.4 AC Testing

AC power frequency tests can be performed only when the cables are of short lengths (1-3 km) and cable capacitance of ~0.25 μF/km. In such a testing a Transformer capable of supplying ~100 kVA at ~ 50 kV is required. Under such conditions, PD, $C$ and tan $\delta$ of the cable can all be measured using proven test setup.

#### 6.4.2.5 Very Low Frequency Testing

Very Low Frequency (VLF) testing was introduced in the beginning of 1960 for testing generator stator insulation [80,81]. Later this method was extended to test power cables to reduce the power required for energizing long lengths of cables. Typically, the frequency range of a high voltage (~100 kV) VLF set is between 0.01 Hz and 1 Hz. At a frequency of 0.1 Hz, the power requirement during testing can be reduced by a factor of 500 compared to that at power frequency.

Among the methods for producing the VLF waveforms are, electromechanical and electronic means. Dissipation factor and PD measurements can also be performed using modern equipment at VLF. PD measurements are possible only when the test facility is discharge free up to the highest level of voltage for measurement. Due to the way in which the VLF is produced the wave form is not strictly sinusoidal and this fact makes comparison of the dielectric parameters and PD measured at these frequencies with those at power frequency uncertain [82].

Discharge measurement on cables is performed as shown in the Fig. 6.2. The facility in such a system includes both measuring the discharges and locating their position along the cable length.

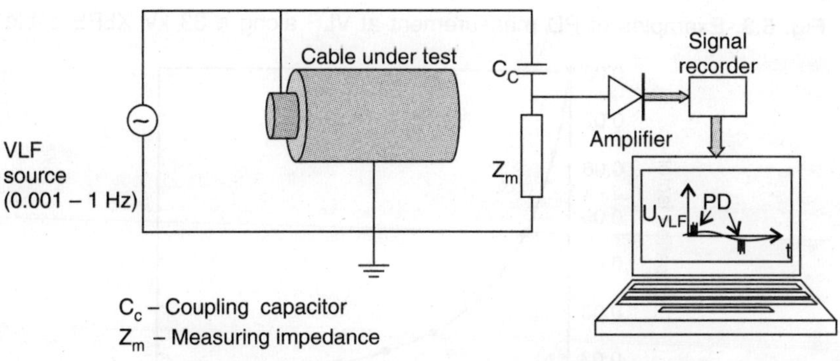


**Fig. 6.2.** VLF partial discharge measuring system

A typical PD pulse display is shown in Fig. 6.3 and Table 6.1 for all the three phases of a 33 kV XLPE cable of about 2.5 km length, including joints. It can be seen from the Fig. 6.3 that the discharge activity at several of the joints along the route length has also been displayed. It is therefore necessary to locate the position of the joints if any by appropriate traveling wave methods so that discrimination between cable discharge and joint discharge can be made.

The capacitance and dissipation factor of a 11 kV XLPE Cable of length 3.14 km, measured at 6.35 kV were respectively 2 μF and 0.08 at 0.1 Hz. The results of the frequency response study of the above cable are shown in Fig. 6.4. It can be seen that there is a change in dissipation factor between 0.1 Hz and 1 Hz by a factor of about 8.

**Table 6.1.** PD parameters of a 33 kV, XLPE cable

| Parameter \ Phase | 1 | 2 | 3 |
|---|---|---|---|
| Inception voltage | 17.5 | 19.05 | 15.8 |
| Maximum discharge | 3141 | 1326 | 2673 |
| Average discharge | 1486 | 639 | 1224 |

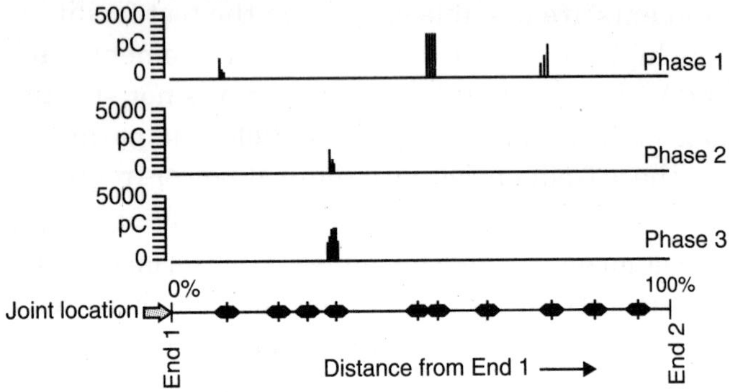


**Fig. 6.3.** Examples of PD measurement at VLF along a 33 kV XLPE cable

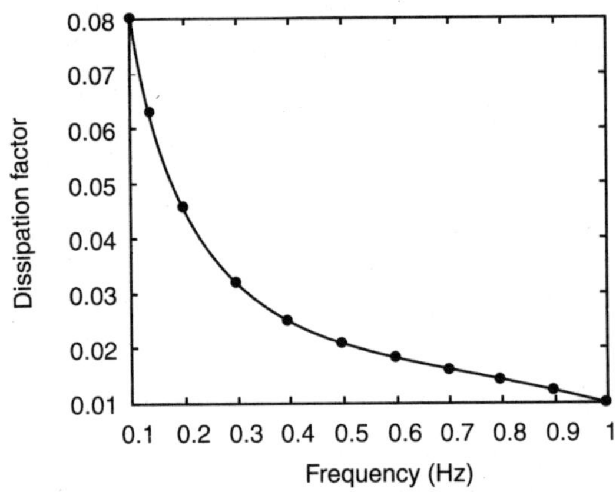


**Fig. 6.4.** VLF dissipation factor (11 kV XLPE cable, 3.14 km; test voltage 6.35 kV)

Although, accurate correlation between the dissipation factor in the VLF range and at 50 Hz is not possible an approximate empirical formula for obtaining equivalent dissipation factor is given as,

$$\tan\delta_f = \tan\delta_{VLF} \times \frac{VLF}{F}^{3/4} \tag{6.2}$$

In a series of measurement on a class of extruded XLPE cables, Whitehead [5] has obtained some important results on energy loss with voltage. Figure 6.5 indicates that a two step increase in PD energy. The change of slope at 4 kV indicates an addition to the dielectric energy caused by PD.

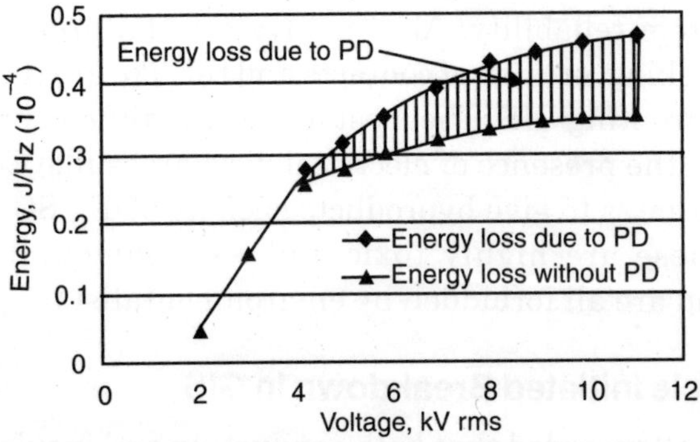


**Fig. 6.5.** Energy loss in a 6.6 kV polyethylene cable due to PD [5]

## 6.5 GAS INSULATED SUBSTATIONS

Gas insulated substations (GIS) or totally enclosed modular substations as they are also called are a major innovation in power transmission and distribution station design. Endowed with proven reliability, and nearly maintenance free operation, as advantages in the medium and high voltage ranges (33 – 400 kV) it has now become a modern practice to install GIS in place of conventional substations. As the name suggests, GIS is a complete enclosure for all energized parts, a metallic encapsulation housing them offers an effective shield against external electromagnetic interference.

Insulating gases comprising $CO_2$, $N_2$ or electronegative gases like sulphur hexafluoride ($SF_6$) are the gases of choice in GIS. The gaseous insulating medium also acts as coolants besides acting as arc quenching medium. The most attractive features of GIS over traditional substations are,

i. Compactness and modularity combined with layout flexibility.
ii. Minimal space requirement and hence cost effective system at very high voltages.
iii. Immunity from adverse environmental electromagnetic interference.
iv. Possibility of mobile outdoor or indoor installation.
v. Possibility of factory assembled modular elements facilitating quick and easy construction.
vi Aesthetically, acceptable layouts.

Notwithstanding these advantages there are drawbacks such as difficulty in visual checks of the component parts, and localization of individual defective elements, relatively longer repair and down times affecting system reliability. Also, inadvertent failures due to support flashover, tracking, erosion of contacts and the like are the certain other problems confronting GIS. The most important difficulty with $SF_6$ filled GIS is that in the presence of electrical discharges like Corona and PD the $SF_6$ dissociates to give byproducts such as $SOF_4$, $SOF_2$, $SO_2F_2$, HF and $SO_2$. These are highly toxic and environmentally hazardous chemicals and are all forbidden by environmentalists.

### 6.5.1 Particle Initiated Breakdown in GIS

It has been well recorded that both conducting and insulating particles which are either already present or generated due to switching operations pose difficult situations by way of partial or complete breakdown of the gas phase. Conducting particles of various shapes sizes and severity, interact with the electric field in GIS whereby, the following deleterious effects are caused.

i. Corona or Partial discharges from the particle depending upon its location.
ii. Partial/Complete breakdown of the gas medium.
iii. When conducting particles are present on the spacers local overheating and tracking occur.
iv. If the particle number density is very large and when they are in the highly stressed regions there is a strong possibility of their bombarding high voltage conductors or support insulators thereby causing surface damage.
v. Due to the activity (iv) above the conductor surface becomes rough and issues more corona.

It is seen that infestation of particles and corona degradation are very closely linked with each other.

It may be mentioned here that loss tangent tip-up can also be used as a diagnostic tool if measurements can be performed at constant frequency and at voltages $U_N$ and about 1.3 $U_N$. As far as the condition monitoring and diagnosis is concerned no uniform standard methods can be prescribed since the indices for diagnosis are almost always the dissipation factor and PD. These two parameters in general do not bear a significant correlation with dielectric strength, a property used as a reference Investigations conducted on a model GIS in the Authors laboratory, were directed towards understanding the mechanisms of degradation therein. The study included among other things the effect of conducting particles (copper and aluminum) of various shapes and sizes positioned at different locations inside the chamber. The chamber was filled with $SF_6$, $N_2$ and atmospheric air at different pressures ranging from 0.1 to 0.6 Mpa. These experiments were conducted with a view to study PD spectrum in GIS and make diagnosis. The electric field required to set a conducting particle in motion, called the lift-off field, $E_L$ can be approximately estimated using the following formulae, assuming that the particle shape is a general cylinder. The lift-off field for a particle supported by a external structure depends on the geometry of the structure itself. By approximating it as a half cylinder, as shown in Fig. 6.6 the estimated lift-off field can be deduced as[83]:

$$E_L = \sqrt{\frac{3\pi a\rho g}{16\varepsilon_0\varepsilon_r}} \tag{6.3}$$

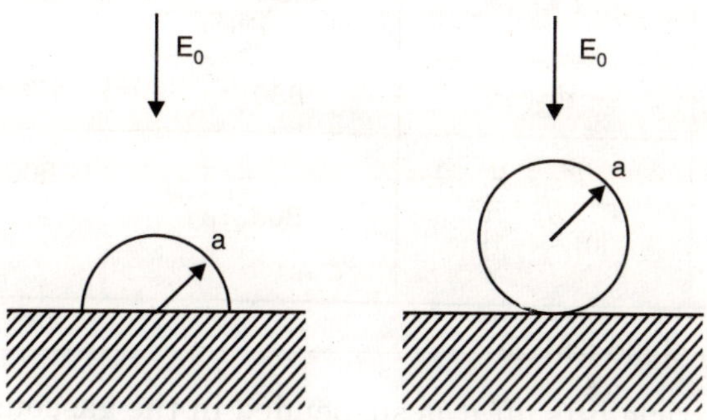


**Fig. 6.6.** Calculation of lift-off field in half cylinder and full cylinder particle geometry

where ρ is the density of a particle, $g$ is the acceleration due to gravity, and $a$ is the radius of the particle.

For a full cylinder of radius $a$ lying on the ground plane, Fig. 6.6, the lift-off force is shown to be as follows [84]:

$$F_L = \frac{1}{3}\pi\varepsilon_0\varepsilon_r E_L^2 a\left(\frac{\pi^2}{3}+1\right) \tag{6.4}$$

Using this expression for force, the lift-off field strength of the particle becomes,

$$E_L = \sqrt{\frac{3a\rho g}{\varepsilon_0\varepsilon_r\left(\frac{\pi^2}{3}+1\right)}} \tag{6.5}$$

Particle initiated breakdowns are among the more important causes for disruption in GIS. A conducting particle in an electric field lifts off at an electric field $E_L$ from the position of rest and travels to the high voltage conductor overcoming the force of gravity, ignoring the effect of viscous drag on the particle. The effective force due to $E_L$ can be calculated when once the amount of surface charge on the particle is known. The Table 6.2 includes the induced charge $Q$, $E_L$ and force $F$ on the conducting particles of different shapes. Using these expressions it is possible to calculate these quantities for any known shape.

**Table 6.2** Expressions for Q, $E_L$ and F for different particle shapes

| Shape of the particle | $Q$ | $E_L$ (kV/m) | $F$ (kN/m) |
|---|---|---|---|
| Spherical | $4\pi r^2 \varepsilon_0 \varepsilon_r E_L$ | $0.5\sqrt{\frac{\rho g r}{\varepsilon_0 \varepsilon_r}}$ | $2\pi r^2 \sqrt{\varepsilon_0 \varepsilon_r \rho g r}$ |
| Cylindrical | $2\pi r l \varepsilon_0 \varepsilon_r E_L$ | $880\sqrt{r\rho}$ | $1760\pi r l \varepsilon_0 \varepsilon_r \sqrt{r\rho}$ |
| Semi-cylindrical | $\frac{\pi l^2 \varepsilon_0 \varepsilon_r E_L}{\ln\left(\frac{2l}{r}-1\right)}$ | $808\sqrt{r\rho}$ | $\frac{808\pi l^2 \varepsilon_0 \varepsilon_r \sqrt{r\rho}}{\ln\left(\frac{2l}{r}-1\right)}$ |

$l$ = the length of the particle

When the particle lifts off it is suspended in the gaseous medium and its further motion depends upon the gas pressure, the mass of the particle and its size.

A point of greater interest is that a conducting particle in an electric field is the primary cause of corona. The magnitude of the corona charge depends on the electric field seen by the particle and hence its position in space. After a critical charging level is reached, charges are dissipated particularly through its tips in the form of electrical discharges, in this case, corona. Now the charge on the particle and hence the electric field are reduced thereby the particle moves away from the central conductor to a region where the equilibrium is established. The same process takes over again and the particle starts moving into a region of higher electric field.

The dynamics of a particle under the action of the electric force and the force due to gravity is shown in Fig. 6.7 as a function of time.

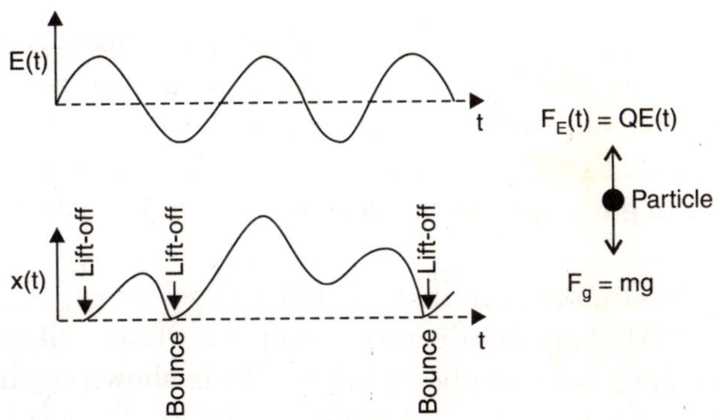


**Fig. 6.7.** Dynamics of a conducting particle in an electric field $E(t)$ with corresponding displacement $x(t)$

It can be seen that although the general motion of the particle follows the field there is a strong tendency for it to rebound from the enclosure walls and possibly from the conductor surface. The motion is thus highly non-uniform and follows no known pattern. In addition to the corona due to a moving particle described above, corona also occurs due to conducting protrusions from the high voltage bus and the enclosure.

Also particles lodged on the surface of insulating supports give rise to PD. A schematic arrangement to a laboratory scale of a GIS is shown in Fig. 6.8 where, position of conducting defects and contact deformities are shown.

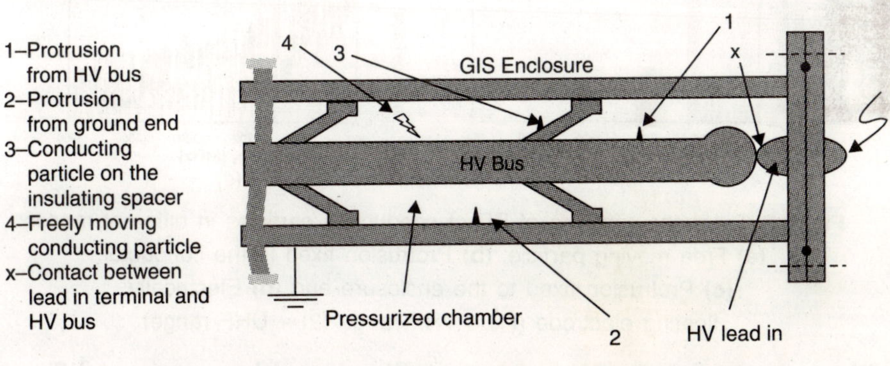


**Fig. 6.8.** A schematic diagram of a section of a GIS showing the position of conducting particle

### 6.5.2 Diagnostic Tests for GIS

It is therefore easy to see that the DTCM for a GIS is mainly indexed to the partial discharges, their shape, frequency content and their spread in the frequency domain.

Onsite testing of GIS for PD and associated effects is done using a UHF/VHF spectrum analyzer which detects and analyzes in frequency domain the PD signals to characterize a possible fault. Other equipment such as wideband PD detector and acoustic emission sensing equipment are also used to get complete information on the PD activity inside the GIS [85].

It is interesting to note that the PD analysis is capable of not only detecting the PD but also localizing the source of PD as well as its position. An intensive diagnostic study on PD in GIS is shown in the Fig. 6.9.

The form and content of PD magnitudes and the frequency components reveal whether the particles are fixed or moving also any ungrounded object in the vicinity of the GIS would produce an effect similar to the other modes.

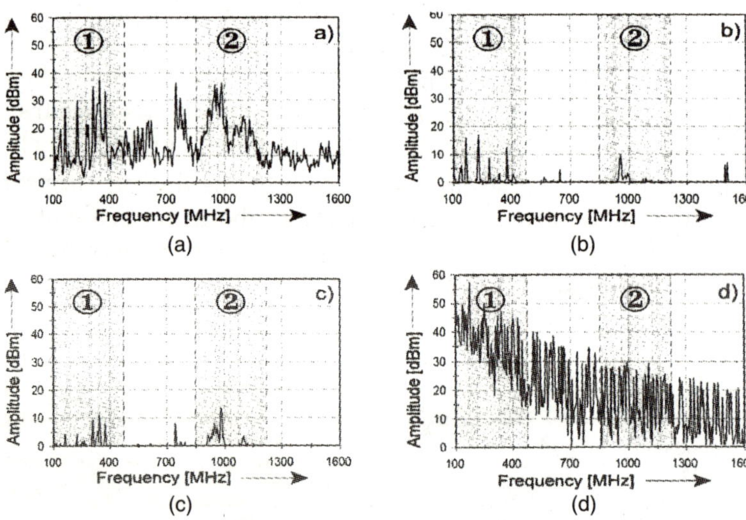


**Fig. 6.9.** Frequency spectra of PD of conducting particles in different modes: **(a)** Free moving particle, **(b)** Protrusion fixed to the conductor, **(c)** Protrusion fixed to the enclosure and **(d)** Electrically floating electrode ((1) – VHF range (2) – UHF range)

The measurement itself involves installing capacitive sensors at different selected locations along the GIS and positioned inside a flanged recess. A PD occurring anywhere in the GIS chamber gives out electromagnetic radiation at their characteristic frequencies. Both electric and magnetic waves travelling in the Transverse Electric and Transverse Magnetic mode interact with the sensors and give outputs at different frequencies (VHF/UHF) which can be detected and analyzed using a spectrum analyzer [86]. The Figs. 6.9 (*a*) - (*d*) are in fact derived from such signals at different sensor locations. Identifying the position of a defect in a

GIS is considerably simpler than in other power equipment. A supplementary diagnostic testing for PD using acoustic emission sensing greatly helps localizing the source of PD.

## 6.6 DRY TYPE TRANSFORMERS

Dry type transformers use class F (Maximum Operating Temperature 150° C) /class H (Maximum Operating Temperature 180°C) insulation for example epoxy bonded glass cloth or mica or quartz powder. The literature suggests that at the present the voltage rating of such transformers are in the range of 1 kV to 45 kV and rated for voltages between a few kVA to about 20 MVA.

Low voltage coils are either Al or Cu in the form of conductor strips and foils. In practice the insulation for High voltage coils is Nomex and inter-disk spacers are, glass fibre impregnated with polyester resin and heat treated. The transformer is either natural cooled or fan cooled depending upon the power rating. Often the transformers are ventilated to improve the cooling efficiency. General requirements for Dry-Type Distribution and Power Transformers are specified, for example, by ANSI/IEEE C57.12.01-89 and others. A cut-away view of a typical dry type transformer is shown in the Fig. 6.10 for information.

The possible defect types normally present in the insulation are gas cavities de-bonded regions between conductor and insulation. The internal mechanical stresses formed during casting and subsequent cooling, if not relieved would give rise to stressed regions which are sources of thermal stress cracking. The faults occurring around defect sites are PD, treeing, and often surface tracking if the external epoxy casting is not maintained well.

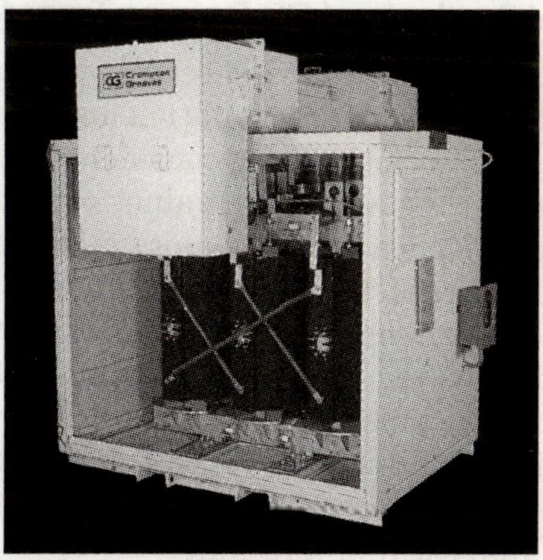

**Fig. 6.10.** An internal view of a Class H insulated dry type transformer

### 6.6.1 Recommended Tests for Epoxy Cast Transformers

The following are some of the recommended tests for epoxy cast transformers:

i. PD measurement involving highest magnitude weighted mean discharge level and PD energy/Hz.
ii. Dissipation factor measurements including the tip up
iii. Polarization index
iv. IR measurements

### 6.6.2 A Case Study

In the following a case study has been reported for a 11kV/440V, 1 MVA, Δ-Y Transformer with insulation of class H type (epoxy-mica). The transformer concerned gave a low partial discharge inception voltage of 4.5 kV, in view of this the utility wanted to investigate the performance of the transformer to ascertain whether or not to continue operating it further.

The system insulation was new and did not show any visible deformity either during manufacture or during subsequent completion of the work. All electrical and other tests suggested by relevant standards were performed and the results there-from did not seem to indicate any inconsistency or incongruence. It was observed, from earlier records, that, dissipation factor measurements were not made before and after heat-run tests, nor during the PD measurements. The measured value of the insulation resistance between windings and between HV and Ground was very high, 20 GΩ and 30 GΩ respectively. In the foregoing analysis, magnitude of partial discharge in excess of about 500 pC becomes the aspect of reference.

### 6.6.3 Method of Analysis of PD Data on Transformers

Data analysis starts with identifying certain parameters of insulation having a direct bearing on the degradation and eventual failure of the insulation in the transformer under consideration. At this point, it is important to note that the numerical values of the PD quantities have only a nominal significance. What is more relevant here is to observe the sequence of discharges, the angular position of their occurrence on the voltage wave, their angular expanse and other connected events. It should be noted, however, that the data examination was initiated in view of the fact that the measured PD magnitudes were found to be far in excess of the expected limits and their dynamical behaviour being decidedly unacceptable. These two aspects constitute the terms of reference for the analytical programme.

PD data analysis is performed based on substantiative response to certain questions from the engineers who performed the PD experiments and the designers of the insulation system and also, on the Author's own assessment of the data presented to him by way of repeating some, or, all the experiments under identical conditions. This *apriori* condition has been met to a large extent in the foregoing treatment. The Author stresses the point, that, the assessment paradigm detailed hereunder, is not a standard method. However, in view of authors earlier research work relating to diagnostic testing, condition monitoring and stochastic failure modeling of epoxy-mica and epoxy glass cloth dry type insulation assessment, there is sufficient reason to believe the veracity of assessment exercise.

Of the many technical parameters required in the analysis, the following are found to be relevant.

1. Are all pulses in a train greater than the permissible value?
2. Is the magnitude of PD nearly constant with reference to time, over, a duration of about an hour at 110% of Normal Operating Stress (NOS)?
3. Same as above but, instead of PD magnitude consider the number density of discharges as the parameter.
4. The extent of the angular spread, $\theta$ of the discharges at inception, is $\theta > 45°$ ?
5. Is $U_i < 1.1U_o$ ?
6. Is the PD magnitude and/or increasing with voltage just beyond $U_i$?
7. Is the value of the dissipation factor tip up greater than 130%?
8. Is the incremental PD energy per pulse corresponding to 7 above.

These parameters have a graded accentuation in determining the long term performance of the transformer. Experience and research results obtained on similar insulation system suggest that these parameters can be given varying weights, $w_i$ depending upon their propensities to age the transformer insulation rapidly. Also, by carefully considering the qualitative, but, definite answers to the above questionnaire a Grade Point, $G_i$, is assigned to each of the entries. The grade points and weightages are given, normally, on a scale of 1 to 8. However, other schemes of gradation are not precluded.

A grading table (Table 6.3) so constructed is included. It was seen from the previous records that a measurement of dissipation factor at $U_i$ was not performed. Actually, the parameters of PD indexed degradation model are sensitive functions of loss tangent at $U_i$. In order to fix the weight for this property, it was found essential to make an estimate of tan δ. This calculation has been performed based on the PD energy/Hz at $U_i$.

Fixation of weights is a involved procedure which calls for experience in handling PD data as well as a good understanding of the processes leading to the insulation degradation due to PD and precludes elaboration here.

Based on the entries in Table 6.3, a figure of Merit, $M$, can be computed as a product of $W_i$ and $G_i$. An Acceptance- Rejection or, a Fail-Pass decision is made thus:

For all values of $M$, such that:

i. $0.9 \le M \le 1$ Excellent

ii. $0.70 \le M \le 0.90$ Good

iii. $0.60 \le M \le 0.70$ Fair

iv. $M \le 0.60$ Bad

v. $M = 0$ Failure

These expressions are only qualitative and serve to classify the insulation based on PD measurements only. If some other dielectric property is used as index of degradation, then the figure of merit, $M$, has to be redefined.

**Table 6.3.** Grading and Parameter rank enumeration

| Guide Number $G_i$ / Parameter $P_i$ | 1 | 2 | 3 | 4 | 5 | 6 | 7 | 8 | Weight function $W_i$ |
|---|---|---|---|---|---|---|---|---|---|
| 1 | ⊗ | | | | | | | | 0.05 |
| 2 | | ⊗ | | | | | | | 0.06 |
| 3 | | | ⊗ | | | | | | 0.08 |
| 4 | | | | ⊗ | | | | | 0.08 |
| 5 | | | | | ⊗ | | | | 0.08 |
| 6 | | | | | | ⊗ | | | 0.15 |
| 7 | | | | | | | ⊗ | | 0.20 |
| 8 | | | | | | | | ⊗ | 0.30 |

### 6.6.4 Estimation of Possible Service Life of Transformers

As an offshoot of this analysis, an estimate of the possible service life of the transformer can be made [78]. Based on the premise that the eventual failure of insulation is predominantly due to PD and the other constituent degrading agent, the maximum temperature of operation, only acts as a catalyst promoting an Arrhenius type degradation process. The model invokes the estimation of life based on IPL with the introduction of appropriate correction factor $g(T, E)$ to take account of the temperature the IPM is rewritten in terms of the PD initiation voltage $U_i$ and normal operating voltage $U_0$.

Based on the Author's analysis on laboratory samples, the PD magnitudes permissible in such transformers can be in the range of 50 and 100 pC and the discharge inception voltage shall be at least 20 % greater than the Normal Operating Voltage (NOV). In the present case there were discharges of magnitude greater than 3000 pC, far in excess of the expected values. It was also observed that the PD inception voltage, as has been mentioned was about 4.5 kV.

$$M = \frac{1}{\sum_i G_i W_i} \tag{6.6}$$

$$(\tan\delta)_{PD} = 4.5\times10^{-4}\times\frac{q_{\text{total}}}{C_x U_i} \tag{6.7}$$

where, $C$ is the capacitance of the specimen in pF, measured at the terminal, $q_{\text{total}}$ is the total charge due to all cavities in pC. Actually, this empirical relation for partial discharge aided dissipation factor and hence the Empirical relationships for partial discharge energy, $E$ and a possible increase of the dissipation factor with the partial discharge magnitude has been proposed by Mole [86], Mason [87], Kruger [7] and others. These relations are given here for reference below.

$$E \approx \frac{1}{\sqrt{2}}\hat{q}\ U_i \tag{6.8}$$

$$(\tan\delta)_{pd} \approx \frac{\sqrt{2}\hat{q}}{\pi\ C\ U} \tag{6.9}$$

where, $E$ is the energy due to integrated charge per hertz, $\hat{q}$ is the weighted mean discharge magnitude (pC) and $U_i$ is the partial discharge inception voltage in kV.

In the present case $q_{\text{total}} = n\ q_{\text{mean}}$ where $n$ is the number of discharges in both the half cycles, taken as about 60 and the mean PD magnitude was about 100 pC so that, $q_{\text{total}}$ works out to 6000 pC. Substituting these values tan δ works out to about 0.04%. It should be remembered that the tan δ so calculated is a component due to the existence of PD.

There are two ways of estimating the life of this type of insulation, the first method is based on temperature corrected inverse power law and the second is based on diagnostic test result and in the present case PD. These two calculations are indicated in the following.

#### 6.6.4.1 Estimation Based on IPM

The general formula for life with a correction factor for the operating temperature is given by [88, 89],

$$L = L_o\left(\frac{1.1 \times U_N}{U_i}\right)^{-n} .g(T,E) \tag{6.10}$$

$$L = L_o\left(1.1 \times \frac{6.3}{4.5}\right)^{-8.5} .g(T,E) \tag{6.11}$$

where, $g(T,E)$ is given by the empirical formula based on the thermal acceleration as:

$$g(T,E) = \left(\frac{\theta_{\max} - \theta_a}{\theta_N - \theta_a}\right)^7 = \left(\frac{155 - 40}{135 - 40}\right)^7 \tag{6.12}$$

In Eqn. (6.12) the exponent has been adjusted to account for a nearly half life for every 10°C temperature increase or decrease. Using these values the estimated life $L$ corresponding to an expected life of 25 years under normal operation, works out to:

$$L = 1.13 \text{ years} \tag{6.13}$$

#### 6.6.4.2 Estimation Based on Diagnostic Measurement (PD)

The Figure of Merit $M$ using the entries in the Table 6.3,

$$M = \frac{1}{\sum_i G_i W_i} = 0.1716 \tag{6.14}$$

The expected life is given by:

$$L = M \times L_0 = 0.1716 \times 25 \tag{6.15}$$

$$L = 4.3 \text{ years} \tag{6.16}$$

It can be seen that the estimated life from these two computation methods are considerably different. However, since the second calculation is based on the diagnostic property of the transformer in question, it is likely that this value is more appropriate but for the sake of higher reliability, the first value may be accepted as it is based on the destructive properties.

## 6.7 HIGH VOLTAGE ROTATING MACHINE STATOR INSULATION

The ageing of rotating machine insulation is somewhat different from those in other power equipment for the reason that the insulation is subjected simultaneously to three stresses, thermal, electrical and mechanical (vibration). Also, thermo-mechanical forces due to differential thermal expansion of conductor and insulation thereon needs to be considered. The conductor insulation of choice in rotating machines is either epoxy-bonded mica or glass cloth. The effect of vibration and thermal stress is to cause de-flaking of mica and de-bonding of glass cloth from the conductor. The electrical stress gives rise to PD and often treeing in the insulation matrix. Over a long duration of operation there is a strong possibility of deterioration of the epoxy component for example, stress cracking and carbonization. This effect further reduces the mechanical stress and the week zones so created are regions of further tree propagation.

The discussion on this topic begins with a detailed description of laboratory controlled ageing experiments on a reasonably large sample size. The test specimens used in the experimental runs are either full size coils or sections thereof as shall be explained later.

A series of experiments on machine coils under electrical thermal and mechanical stresses have been performed when the stresses were applied individually and in combination. A test setup for this purpose was developed in the laboratory for testing full coils as well as coil sections as shown in the Fig. 6.11.

As regards mechanical testing of coils, it is difficult to quantify the vibration induced deterioration in a straight forward manner. Laboratory controlled experiments on rotating machine coils under conditions similar to actual rotating machines were performed [90, 91] on machine coils rated for 11 kV among the characteristics of the coil studied are effect of the amplitude of vibrations, the natural frequencies of the coils and the time varying forces applied to the end of the coil. The overhang position of the coil was simulated.

The cross sections of a 11 kV and a 6.6 kV rotating machine stator conductor bars are shown in Fig. 6.11. A very important point to be noted is that the applied electric stress on the conductors is referred to that stress appearing on the insulation adjacent to the flat portion (indicated as A in Fig. 6.11 (*c*)) of the conductor. Actually the true maximum stress occurs at the corners of the conductor (indicated B in Fig. 6.11 (*c*)). The conductor corners are appropriately rounded off to relieve the stresses in this region. Depending upon the corner radius and the thickness of the insulation, the electric field at B can be as high as 5-10 times the electric field at A [17]. The Table 6.4 gives approximate maximum stresses on the conductor insulation for different rated voltages and insulation thickness for two corner radii of each class.

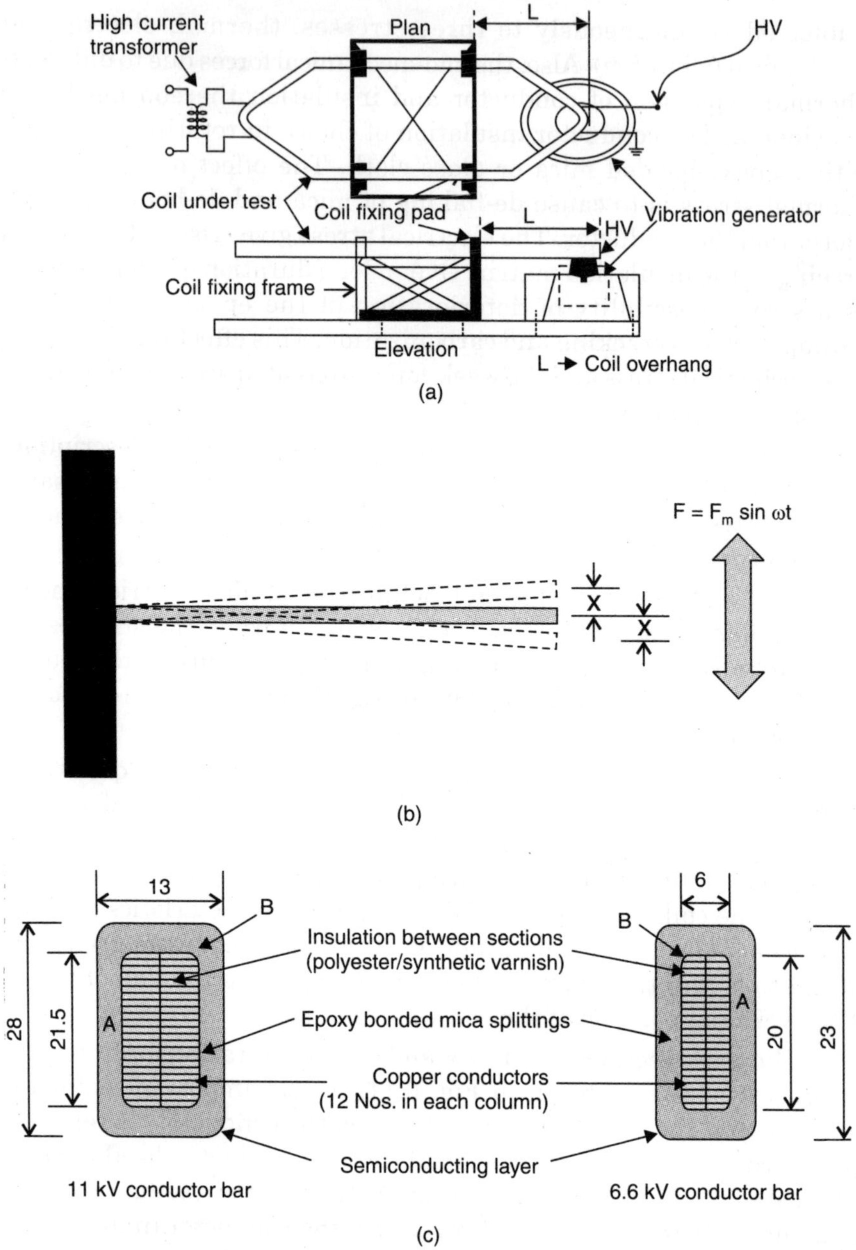


**Fig. 6.11.** Complete test facility for ageing of rotating machine stator coils **(a)** A layout of the full coil under test, **(b)** Diagram showing the vibration of the coil over hang and **(c)** Cross-section of machine coils tested

**Table 6.4.** Electric field strength for various insulation thickness and edge radii

| Rated voltage $V_M$ kV | Insulation thickness d mm | Electric field strength (Uniform) E, kV/mm | Electric field strength (Edge) $E_{max}$ kV/mm | Ratio $(E_{max})/E$ | Conductor edge radius $r_0$ mm |
|---|---|---|---|---|---|
| 10 | 3.0 | 1.92 | 5.37 | 2.79 | 0.6 |
| 10 | 3.0 | 1.92 | 4.16 | 2.16 | 1.0 |
| 21 | 6.0 | 2.02 | 8.42 | 4.17 | 0.6 |
| 21 | 6.0 | 2.02 | 3.67 | 1.82 | 3.0 |
| 27 | 7.3 | 2.13 | 10.07 | 4.72 | 0.6 |
| 27 | 6.5 | 2.40 | 4.51 | 1.88 | 3.0 |
| 30 | 8.0 | 2.16 | 10.84 | 5.02 | 0.6 |
| 30 | 7.2 | 2.41 | 4.72 | 1.96 | 3.0 |
| 40 | 9.5 | 2.43 | 5.39 | 2.22 | 3.0 |
| 40 | 8.0 | 2.89 | 4.83 | 1.67 | 5.0 |

$$E_{max} = E\left(\frac{d}{r_0 \ln\left(\frac{d+r_0}{r_0}\right)}\right)$$; Electric field strength at the edge of conductors

$$E = \frac{V_N}{d\sqrt{3}}$$; Electric field strength of the uniform field

It can be seen that long duration electrical ageing in machine insulation being a point process it is only the maximum stress that counts in deciding the probable life of the machine. Referring to Fig. 6.12, coil sections, included between tightly packed metal barriers to simulate the slots of the rotating machine so as to reproduce the actual conditions of the stator were used to increase the sample size. A large number of experiments under single and combined accelerated electrical ($E$), mechanical ($M$) and thermal ($T$) stresses have been run to acquire failure data. The coil sections were intended to reduce cost of running ageing experiments while full coils had to be used in experiments involving vibration (mechanical) studies. Dielectric and other measurements including breakdown tests were performed at different intervals of time to record the dynamics of degradation of stator coils. The scheme of testing is presented in the Tables 6.5 & 6.6 where the stresses and their combination are indicated along with their magnitudes.

The coils were aged at different accelerated electrical thermal and mechanical stress levels for extended duration of times up to about 3000 hrs. Figures 6.13 and 6.14 show the times to failure under the stated conditions have been plotted on the corresponding probability papers. At least 10 specimens have been put to test at each stress. It may be

**Table 6.5.** Single and combined stress test schemes

| Stresses applied \ Scheme of application | E | T | M | E+T * | E+T+M * | Remarks |
|---|---|---|---|---|---|---|
| E (kV/mm) | 20 | - | - | 20 | 20 | *In combined stress application all combinations of stresses indicated were applied |
| | 22 | - | - | 22 | 22 | |
| | 24 | - | - | 24 | 24 | |
| | 26 | - | - | 26 | 26 | |
| | | | | + | + | |
| T (°C) | - | 180 | - | 155 | 155 | |
| | - | 210 | - | 180 | 180 | |
| | - | 230 | - | 210 | 210 | |
| | | | | | + | |
| M (N/cm²) at 100 Hz | - | - | 44.2 | - | 44.2 | |
| | - | - | 55.2 | - | 55.2 | |
| | - | - | 66.4 | - | 66.4 | |

noted that, the 63.2th and 50th percentiles respectively correspond to the nominal lives on the Weibull and the Log-Normal plots respectively. It may also be observed that, on the Log-Normal plot, all the specimens did not fail in a time limit of about 3000 h.

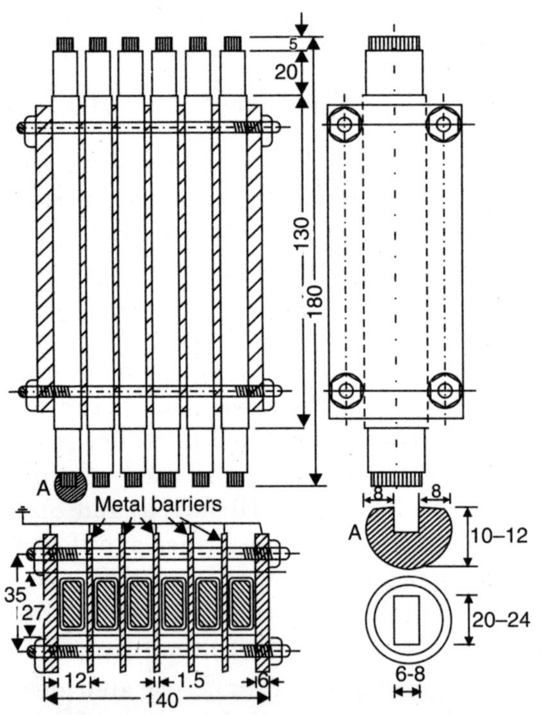


A–Corona suppressing electrodes

**Fig. 6.12.** Assembly of rotating machine coil sections for T+ E ageing

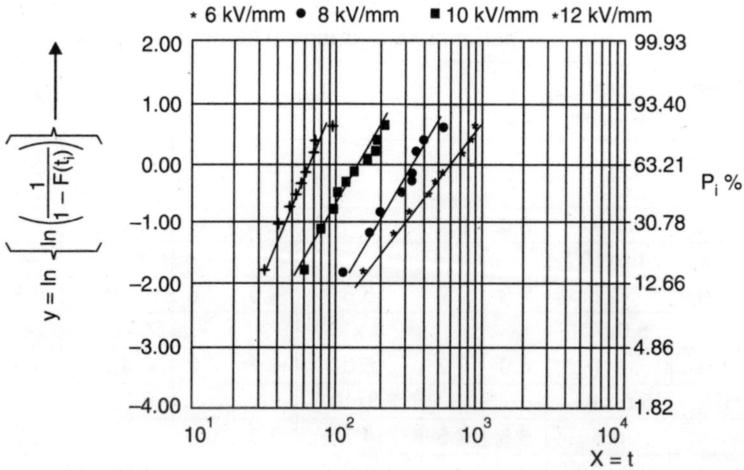


**Fig. 6.13.** Weibull plots of times to failure of rotating machines ( T= 210° C)

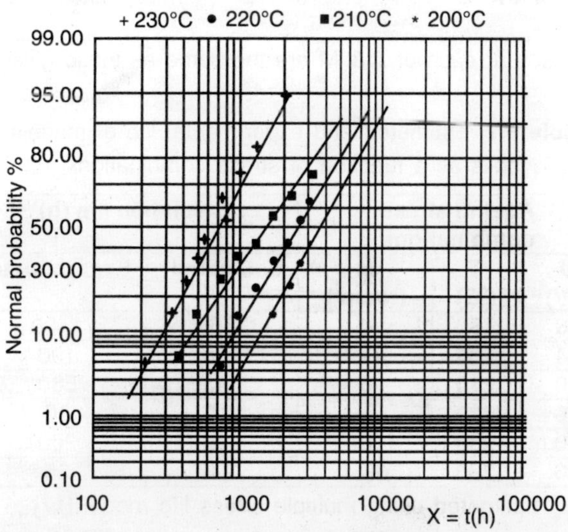


**Fig. 6.14.** Log-Normal probability plot of times to failure of rotating machine insulation ($E$ = 0)

The Table 6.6 shows the Inverse Power law (IPM) parameters with their 95% confidence intervals (CI). These computations have been performed using analytical techniques of choice and in the present case maximum likelihood method. Since in some cases all the specimens put to test have not failed over a time limit of 3000 h, censored data analysis had to be performed. The estimated life of the generator coils at different temperature and other stresses arc tabulated in Table 6.7. Similar

**Table 6.6.** Power law parameters and their 0.95 confidence intervals*

| M → ($N/cm^2$) | | 0 | | | 55.2 | | | 66.4 | | |
|---|---|---|---|---|---|---|---|---|---|---|
| T (°C) | P | L | PE | U | L | PE | U | L | PE | U |
| 155 | n | 7.9 | 8.4 | 8.9 | - | - | - | - | - | - |
| | k | $4.0\times10^{14}$ | - | - | - | - | - | - | - | - |
| 180 | n | 6.7 | 7.3 | 7.9 | 5.6 | 6.6 | 7.6 | 5.1 | 6.3 | 7.6 |
| | k | $5.0\times10^{12}$ | - | - | $2.4\times10^{11}$ | - | - | $7.4\times10^{10}$ | - | - |
| 210 | n | 6.1 | 6.9 | 7.7 | 5.0 | 6.1 | 7.2 | 4.6 | 5.8 | 7.0 |
| | k | $6.0\times10^{11}$ | - | - | $3.3\times10^{10}$ | - | - | $8.8\times10^{9}$ | - | - |
| 230 | n | 5.5 | 6.5 | 7.5 | 4.7 | 5.8 | 6.9 | 4.3 | 5.5 | 6.7 |
| | k | $6.0\times10^{10}$ | - | - | $4.6\times10^{9}$ | - | - | $1.5\times10^{9}$ | - | - |
| P-Parameters of IPM, L – Lower 0.95 CI, PE – Point estimate, U– Upper 0.95CI | | | | | | | | | | |

* Electrical stresses at each of T & M are the same as those given in Table 6.5.

**Table 6.7.** Estimated and experimental life of machine coils as a function of stress combinations

| Ageing stress combinations | | | Insulation life (h) | |
|---|---|---|---|---|
| U (kV) | T (°C) | M ($N/cm^2$) | Estimated* | Experimental |
| 20 | 180 | 0 | 1556 | 1400 |
| 24 | 180 | 55.4 | 186 | 190 |
| 20 | 210 | 0 | 643 | 680 |
| 26 | 210 | 55.4 | 77 | 80 |
| 20 | 230 | 0 | 290 | 200 |
| 26 | 230 | 66.4 | 24 | 25 |

*Estimated using multiple stress life model [17]

computations obtained using the combined stress life models have also been included. Based on these accelerated stress test results the expected life at the normal operating stresses has been worked out and is presented in Table 6.7. Many diagnostic properties have been measured along with the breakdown voltage and a correlation has been established between these properties and estimated values of statistical parameters of positive and negative pulse magnitude distribution are at different ageing are intervals given in Tables 6.8 and 6.9.

**Table 6.8.** Correlation coefficients for rotating machine insulation

| Parameter | G | Δ tanδ | ΔC/C | $E_{PD}$ | $R_{dc}$ | $u_d$ | $S_k$ |
|---|---|---|---|---|---|---|---|
| G | - - | 0.4483 | 0.2861 | 0.4692 | – 0.2778 | – 0.9429 | – 0.9472 |
| Δ tanδ | 0.44 | - - | 0.9282 | – 0.2210 | – 0.9556 | 0.6462 | – 0.5725 |
| ΔC/C | 0.2861 | 0.9282 | - - | – 0.5509 | – 0.9943 | – 0.4181 | – 0.3249 |
| $E_{PD}$ | 0.4692 | – 0.2210 | – 0.5509 | - - | 0.4971 | – 0.4845 | – 0.5678 |
| $R_{dc}$ | – 0.2778 | – 0.9556 | – 0.9943 | 0.4971 | - - | 0.4412 | 0.3505 |
| $u_d$ | – 0.9429 | 0.6462 | – 0.4181 | – 0.4845 | 0.4412 | - - | 0.9949 |
| $S_k$ | – 0.9472 | – 0.5725 | – 0.3249 | – 0.5678 | 0.3505 | 0.9949 | - - |

**Table 6.9.** Estimated values of statistical parameters of positive and negative pulse magnitude distribution at different ageing intervals (G=0.28)

| Statistical Parametors | Duration of ageing (h) | | | | | | | |
|---|---|---|---|---|---|---|---|---|
| | 500 | | 1500 | | 2500 | | 3000 | |
| | + | – | + | – | + | – | + | – |
| μ | 3.55 | 5.08 | 6.00 | 6.30 | 7.73 | 6.91 | 9.06 | 7.65 |
| $\sigma^2$ | 1.72 | 3.35 | 3.62 | 5.63 | 5.28 | 5.98 | 3.26 | 3.47 |
| σ | 1.31 | 1.83 | 1.90 | 2.37 | 2.30 | 2.45 | 1.81 | 1.86 |
| $S_K$ | 0.74 | 0.77 | 0.39 | 0.36 | 0.04 | 0.01 | –0.21 | –0.30 |
| K | 3.33 | 2.79 | 2.63 | 2.79 | 2.49 | 2.23 | 2.53 | 3.22 |
| **μ = Weighted mean pulse magnitude (mV); $\sigma^2$= variance $(mV)^2$; K = kurtosis; σ = standard deviation (mV); $S_k$= skewness** | | | | | | | | |

It was observed that a careful statistical treatment of the data based on the central tendencies of the probability density functions of the PD magnitude distribution, Fig. 6.15, it was observed that in particular the skewness ($s_k$) of the distribution is a parameter of particular significance. In the beginning of the ageing cycle, the distribution is skewed to the right ($s_k < 0$) and slowly the coefficient of skewness approaches zero ($s_k = 0$, indicating a symmetric or Gaussian distribution) and later the density function skews to the left ($s_k > 0$). Failure invariably occurs after this condition is reached.

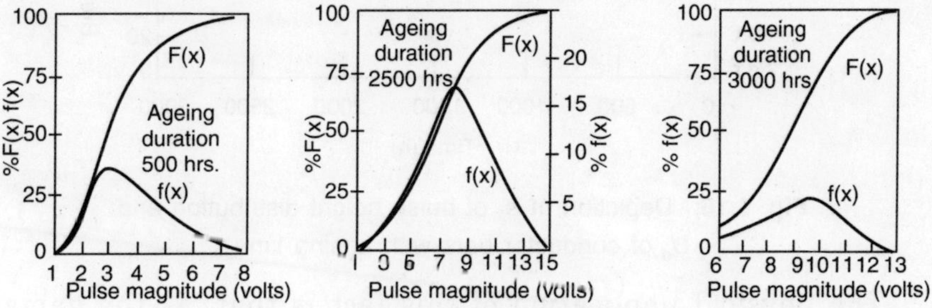


(a) Positive Pulses

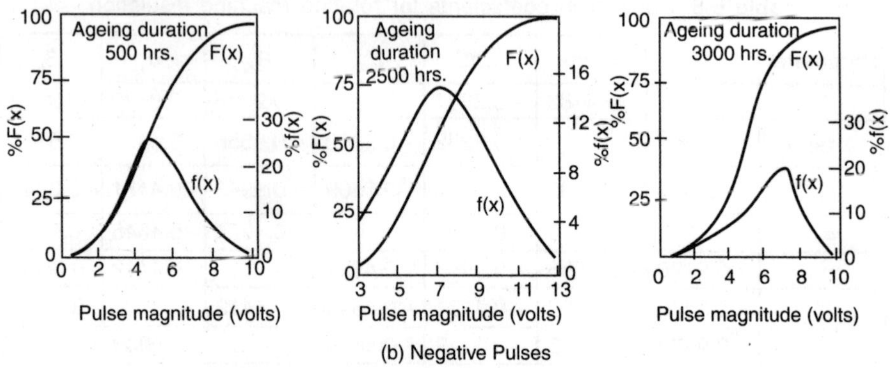


**Fig. 6.15.** Probability distributions of PD pulse magnitudes at different ageing times

Interestingly, the point in time where $s_k$ approaches zero, the breakdown strength of the section of the bar under test would start decreasing very rapidly around this point. This aspect is shown in the Fig. 6.16. This fact is true for both positive and negative PD pulses. However, the scales of magnitudes are different. It was therefore decided to use skewness as a possible parameter for indicating the amount of ageing. The dielectric strength decreases with increased ageing. Therefore a strong correlation exists between the skewness of the density function of pulse distribution and the dielectric strength.

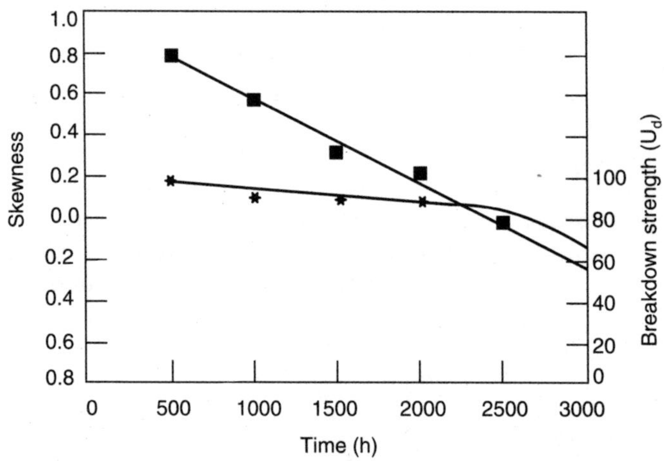


**Fig. 6.16.** Depiction of $s_k$ of pulse height distribution and $U_d$ of conductor bars with ageing time

The physical implication of this fact is that, as the damage accumulates, the number of discharge sites in the body of the insulation such as crevices, tree-like breakdown paths, and micro cracks grows. As a consequence, the number density of discharge pulses increases and the maximum amplitude of discharges also increases. This is borne out by the fact that the weighted mean amplitude as well as the modal value of the distributions can be seen to increase in nearly equal steps.

## 6.7.1 Recommended Tests for HV Rotating Machine Stator

Based on these extensive laboratory design tests and failure models developed earlier it is possible to proceed with the diagnostic testing and condition monitoring of rotating machines in service.

Among the more important functional tests found to be essential in DTCM are:

i. AC/DC withstand/proof tests
ii. Polarization index/insulation resistance measurement
iii. Power frequency AC testing
    a. Dissipation factor measurement
    b. Partial discharge measurement

### *6.7.1.1 AC Withstand Test (AC hipot test)*

The AC test method (IEEE 56-2006) requires application of a voltage 150% of the rated stator voltage, between the conductor and machine frame and maintained for 1 minute. At this increased voltage level the insulation is considerably highly stressed and any gaseous defects either formed during operation or already present are shown up upon a PD measurement. In very large machines, for example a 13.8 kV stator winding the capacitance will be in the range of 1 μF, the power requirement for testing is greater than 150 kVA. Such sources are found to be difficult to transport to the machine site unless one is already available. A DC equivalent test or VLF test (0.1 Hz) [92] in such cases can get over this problem of large power requirement.

### *6.7.1.2 DC Withstand/Proof Test (DC hipot test)*

In a recent recommendation the IEEE 95-2002 suggests application of direct voltages depending upon the machine rating given by about 2U+1 kV where U is the rated r.m.s phase to phase voltage of the stator winding. It is believed that this voltage level corresponds to a phase to ground fault condition. Also under DC since the partial discharges are not very frequent, this test does not age the insulation as much as its AC counterpart. Three ways of applying the DC voltage are suggested in the standards. A generally accepted method of test is one in which a continuously increasing voltage of the calculated levels is applied and maintained for a duration between 1 and 5 minutes. The voltage source is then quickly disconnected and the stator grounded. A sound insulation causes no-tripping of the circuit breaker whereby a transient condition is not experienced. Should there be a tripping of the breaker the winding insulation is deemed to be week and requires a overhauling.

#### 6.7.1.3 Incremental Dissipation Factor and its Tip-up

Figure 6.17 shows, to an arbitrary scale, this aspect. $\Delta \tan \delta$ is sometimes called as the "loss tangent tip-up or simply tip-up", taking *P* as a point of reference. This test is usually made yearly or half yearly to see how rapidly the tip-up is increasing. Similar remarks apply to capacitance tip-up as well. IEC has given some guidelines to decide on whether the insulation has suffered a cognizable deterioration based on the tip-up measurements.

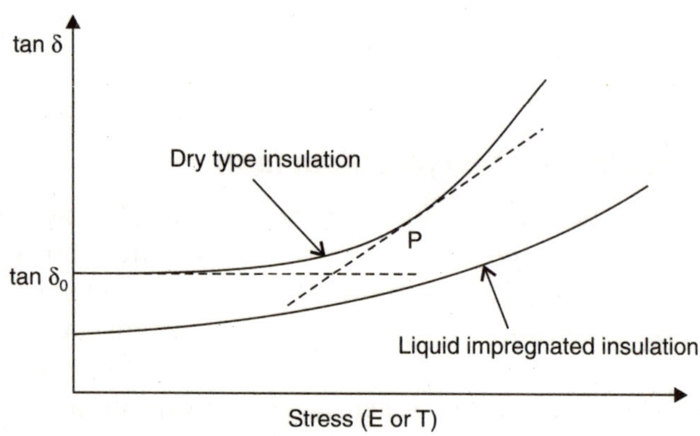


**Fig. 6.17.** Behaviour of dissipation factor with stress or time

A further description may perhaps be required to understand the relationship between internal and surface discharges and the dissipation factor. In dry type insulation, referring to Fig. 6.18, the equivalent circuit includes two additional resistive components $R_{pd}$ and $R_{surface}$. While $R_{pd}$ represents the mean pulsive current due to PD, $R_{surface}$ represents the intermittent current over the surface. In view of the fact that, due to constant occurrence of these two effects, the corresponding components show a decreasing tendency. The PD measurements being made at the terminals these two components can never be separated. An increase in dissipation factor as a function of time is thus due to the following factors:

i. Physical enlargements of existing PD sites.
ii. Opening up of new PD sites.

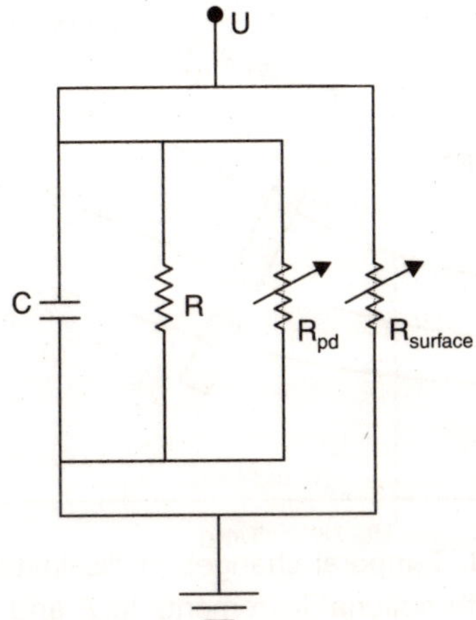


**Fig. 6.18.** Simplified equivalent circuit for representing insulation resistance in the presence of PD

iii. Partial or total carbonization at the discharge locations due to the PD energy.

iv. Continuous surface discharge activity and with a possible increase in number of discharge paths.

All these factors are dependent upon the frequency of the applied field and the temperature of operation.

There are some difficulties in using tan δ and its tip-up as an index of degradation. First of all, the scatter in such measurements is often very large which makes them highly unreliable. Also, the correlation the DF and other properties like dielectric strength, pd quantities and so on is quite poor, except perhaps when the system is approaching end point.

In Liquid Impregnated Systems the dissipation factor is a combined function of the temperature coefficients of the dissipation factor of both solid and liquid phases. Figure 6.17 shows the changes in dissipation factor for such systems as can be seen there from, that, tand changes slowly but uniformly. It is also possible to use as a diagnostic tool the time related changes at constant stress. The Fig. 6.19 suggests that with increase in $E$ or $T$ for a given inspection time ($t_{ins}$), the dissipation factor shows an increasing trend. However, the changes themselves are nearly the same with time. In diagnosis a notional tip-up can also be invoked here by which the incremental dissipation factors at stresses $s_1$, $s_2$ and $s_3$ can be recorded for later decision making.

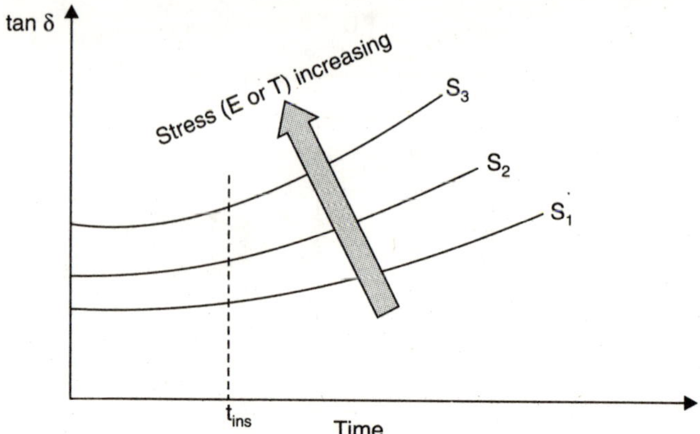


**Fig. 6.19.** Temporal changes in dissipation factor with notional increments in *E* and *T*.

### *6.7.1.4 Partial Discharge Parameters*

A gas filled cavity breaks down due to a concentration of electric stress across it due to the differing relative permittivity of the gas (~1.0) and that of the surrounding solid dielectric medium, ε (> 1.0). Assuming a spherical cavity, the cavity field $E_c$, seen by it, is approximately given by:

$$E_c = \frac{3\varepsilon E}{1+2\varepsilon} \tag{6.17}$$

where, $E$ $(U/t)$ is the uniform field in the dielectric outside the cavity.

The condition for alternating voltage breakdown of very narrow air or gas filled gaps is determined by the phenomenological Paschen's law. According to this law, there exists a minimum (critical) voltage below which the gap cannot breakdown, no matter how thin the gap is. Also, the Paschen's minimum, as this is called, is found to be a function of the product of the cavity thickness normal to the direction of the applied field and the pressure of the gas in the cavity, Fig. 6.20. The discharge inception voltage, $U_i$ is given approximately by:

$$U_i = E_g\left\{t + t_g(\varepsilon - 1)\right\} \tag{6.18}$$

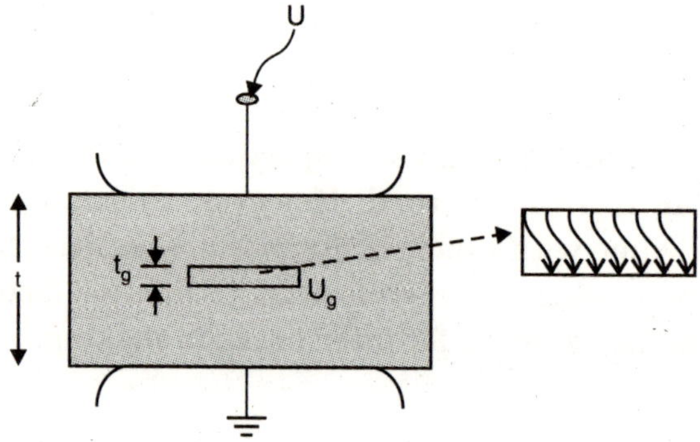


**Fig. 6.20.** Breakdown of a gas cavity enclosed in an insulating slab

where, $E_g$ and $t_g$ are respectively the electric strength and the average thickness of the gas cavity and $t$ is the thickness of unbroken solid insulation shown in Fig. 6.20. $E_g$ is about the same for the gas when it is enclosed between parallel conducting planes having the same spacing as $t_g$. Based on these results, the magnitude of charge transfer, the energy associated with the discharges, and other related parameters can now be computed as under.

Referring to the Fig. 2.13(a), let $C_0$ and $C_1$ be the capacitance of the cavity and the unbroken capacitance of the specimen in series with it. When the conditions in the cavity is such that it is unable to support the Paschen-minimum voltage, it breaks down. A situation like this is simulated by a spark gap or a voltage limited switch, which closes and short circuits the gap at $U = U_{\text{Paschen}}$, which can be identified with $U_i$.

If $\Delta U$ is the impulsive dip in the voltage across the terminals caused by a single discharge, then, $q_0$ the true charge transferred across the cavity, (assuming $C_1 >> C_0$) is given by:

$$q_0 \cong (C_0 + C_1)\Delta U \tag{6.19}$$

It may be observed that while $C_1 \sim C_0$ and $\Delta U$ can be measured by a sensitive detector connected outside the terminals of the specimen, $C_0$ remains unknown. It is for this reason that the magnitude of the true charge cannot be directly measured by any detector. Alternately, to a higher degree of approximation, the displacement, (apparent) charge, $q$, given by:

$$q \cong C\,(\Delta U) \tag{6.20}$$

where, $C$ is the specimen capacitance can be measured with an appropriate detector capable of sensing such small voltages as $\Delta U$. It may be seen that neither $C$ nor $q$ represent the actual situation at the site of the cavity. However, Eqn. (6.20) tends to overestimate the (apparent) charge released and hence the amount of degradation caused to the insulation. This may be deemed as a factor working in favour of insulation diagnosis. Avoiding details, it may be easily shown that energy, $E$ of the discharge is proportional to the product of $q$ and $U_i$ thus:

$$E \approx \frac{1}{2}\frac{C_0 C_1}{C_0 + C_1}(\Delta U)\ U_{i(\max)} \tag{6.21}$$

This can be further simplified to:

$$E \approx \frac{1}{\sqrt{2}} q\ \ U_{i(\max)} \tag{6.22}$$

In an insulation system, there are a large number of defects, where, discharge processes take place at each defect site corresponding to their Paschen's minima. The PD detector responds to the apparent charge transfer at each location. In effect, a large number of discharge pulses of varying magnitudes and repetition rates are recorded (Fig. 3.8 – Fig. 3.10). It may therefore be seen that locating the defect site is quite out of reach. Another important point to be noted here is that, the pulse magnitudes, their shapes and frequencies vary both in space and time. In view of the complexities involved, it is difficult to measure, analyze and interpret PD data.

A point in favour of using PD as a diagnostic tool is that the partial discharge patterns reveal, fairly clearly the nature of defects, their possible position in the insulation matrix (Location of PD source depends on the type of equipment in question). The PD signatures of differing types of faults, their nature and location, are depicted in Fig. 3.8 – Fig. 3.10. It is possible to distinguish between the patterns and identify the defect. However, in large equipment, many of these defects occur simultaneously. In such an event, this information may not be of particular use and location of PD source has a different connotation.

Partial discharge measurement in massive insulation systems, for example, large power transformer, or, a very long length of power cable poses serious problems, especially when the source is diffused and buried deep within the system due to inordinate attenuation of charge signals. Under such conditions, the measured charge magnitudes are inaccurate and give rise to considerable signal distortion. Partial discharges are characterized, in addition to the charge magnitude and pulse shape, by certain other quantities like pulse phase angle, its angular disposition on the applied voltage wave, the distribution of the pulse magnitudes, total energy per Hz and other aspects. It is therefore necessary to record and analyze all these quantities over a considerable duration of time, before drawing meaningful conclusions on the status of insulation.

#### 6.7.1.5 Case Studies on Rotating Machine Rotor and Stator

A generator of rating 12 kV, 12kA, 25 MVA, 3 phase, 50 Hz, was tested on-site for electrical tests after 42 years of operation to assess the insulation condition. Tables 6.10 and 6.11 summarise the test results on Stator and Rotor, respectively. The test results indicate healthy condition of rotor and stator winding. As the machine was 42 years old, it was suggested to thoroughly clean and dry the winding before putting back into service.

**Table 6.10.** Tests results on stator winding

| Test | | Test connection (1) | (2) | (3) | Recommended values | Remarks |
|---|---|---|---|---|---|---|
| IR, MΩ | 60 Sec | 1170 | 1230 | 1060 | | RI & PI values are quite high |
| | 600 Sec | 5440 | 5960 | 5070 | | |
| Polarization index | | 5 | 5 | 5 | >2 | |
| Capacitance, nF at Un | | 0.98 | 0.98 | 0.98 | | Low dielectric loss, ΔC and ΔT indicate low void content. |
| Capacitance tip up, ΔC | | 0.44 | 0.38 | 0.53 | | |
| Tan δ (%) at Un | | 1.6 | 1.8 | 1.9 | | |
| Tan δ tip up, ΔT | | 0.13 | 0.16 | 0.18 | | |
| *PD at phase voltage , pC | | 3000 | 3500 | 2500 | 10,000 | Inception voltage is low, PD magnitude is also low and seems to be in contrast for aged machine insulation. |
| PD inception voltage, kV | | 3.5 | 2.2 | 1.0 | > operating line to ground voltage | |
| PD extinction voltage, kV | | 2.1 | 1.3 | 0.5 | | |
| DC leakage Current μA, at 10 kV dc | | 4.6 | 4.1 | 4.7 | | Gradual increase in leakage current - no weak points in the overhang. |
| Surge comparison test at 5 kV | | No abnormality observed between phases | | | | No inter-turn fault. |
| Stator winding resistance, mΩ at 30°C | | 0.9 | 0.9 | 0.9 | < 5% of commiss-oning test value. | Values are in satisfactory range. |

(1) R versus Y&B shorted and grounded, (2) Y versus B &R shorted and grounded, (3) B versus R&Y shorted and grounded

*Sensitivity of PD measurement was 2000 pC as the background noise was high at the time of measurement.

**Table 6.11.** Tests results on rotor winding

| Test conducted between +ve and –ve slip rings | | Test result |
|---|---|---|
| Rotor winding resistance, mΩ at 30°C | | 158 |
| IR, MΩ | 15 Seconds | 2.5 |
| | 60 Seconds | 2.5 |
| Recurring Surge Oscillograph test | | Stable & identical patterns and no abnormalities. |

## 6.8 INDIRECT ELECTRICAL DIAGNOSIS

### 6.8.1 General

In large generators and motors, with fully insulated stator windings and grounded iron slots, it is safe and convenient to energize the winding and simultaneously use an inductively coupled detector probe across the slots of the stator. Capacitively coupled detector probes have also been used, placed against the insulated coils usually outside the iron on the end turns. The latter is less safe and less accurate because of a difficulty in reproducing the capacitance coupling. With such probes, the problems and uncertainties of terminal measurements are reduced. The coupling to the discharge site can be as close as one shot out of many. This reduces the amount of superposition of many pulses and the amount of attenuation between the discharge site and the detector. As will be elaborated in the next paragraphs, this technique can often isolate a larger defect to one coil in one slot.

### 6.8.2 Electromagnetic Probe Measurement of Discharges in Large Generator and Motor Stators

An inductively coupled probe is illustrated in Fig. 6.21 was suggested by Dakin et al. [93] as a pick up for discharges from rotating machine stator coil. A soft iron finite-core around which is wound a small pick-up winding is placed across the slot iron. Either a narrow- or wide-band pulse detector-amplifier and quasi-peak output meter is connected to the coil through a coaxial cable. Originally, measurements have been made with a narrow-band radio noise meter tuned to the resonant frequency of the coil and attached cable and detector capacitance, and this works very well. The results of early tests with a megahertz probe on an old water wheel generator are shown in Fig. 6.22. It is obvious that slots with large discharge defects are clearly identified. Using such a probe with an injected calibrating pulse, capacitively coupled with an adhesive metal tape attached to the surface of an insulated coil on the

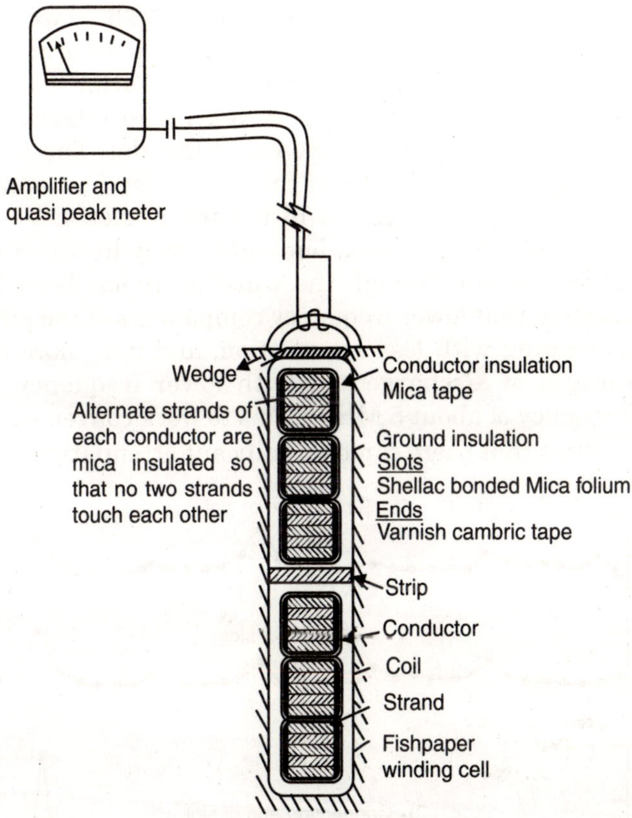


**Fig. 6.21.** Partial display inductive pick-up detector probe for generator windings [93]

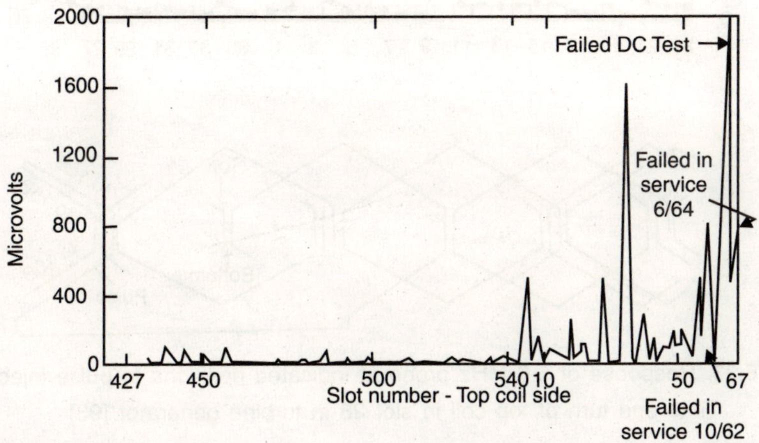


**Fig. 6.22.** 5MHz corona probe tests on 38 year old water wheel generator, 5 kV to ground [93]

end turn of a turbine generator, the results of Fig. 6.23 were obtained. This diagram shows quite clearly the way in which a pulse at one location goes around and is attenuated in the winding, appearing with maximum intensity at the closest slot, 28, and at next greatest intensity at slot 10 when a connected coil appears in the top of the slot. The pulse is also capacitively coupled to the bottom coil of slot 28 and appears with significant level where that coil is in the top of slot 4. The diagram illustrates quite well the complex, but logical, way in which individual discharge pulses spread through the winding. It has been found , as might by expected, that lower frequency components of the pulse spread through the winding with less attenuation, and it is more difficult to isolate the origin of discharge site with lower frequency detectors. Detection frequency of about 5 MHz seems to work conveniently. Figure 6.23 also indicates that there can be significant attenuation along a slot.

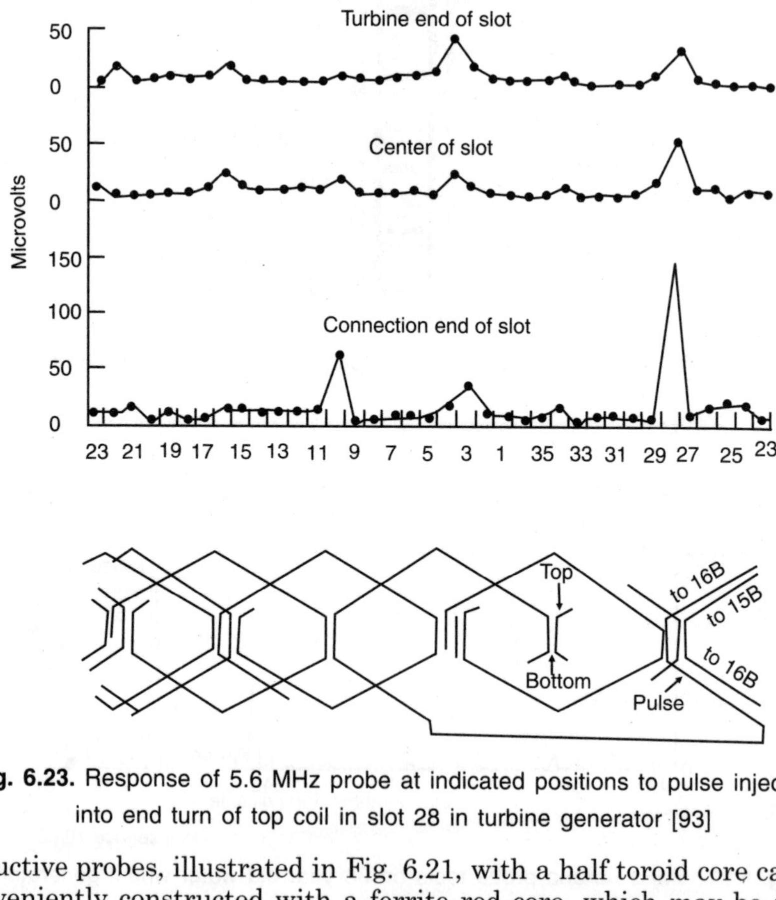


**Fig. 6.23.** Response of 5.6 MHz probe at indicated positions to pulse injected into end turn of top coil in slot 28 in turbine generator [93]

Inductive probes, illustrated in Fig. 6.21, with a half toroid core can be conveniently constructed with a ferrite rod core, which may be more convenient for use on different width slots. These probes are remarkably insensitive to random electromagnetic noise and have been used without problems in electrically noisy factories and generating stations. They can be calibrated in pico-Coulombs by placing a small loop of wire in series with a calibrating capacitor of 10pF and a step-voltage pulser, within the arc of a half-toroid and the probe. To simulate better the conditions in the application, the loop of wire from the calibrator can be laid close the top of a coil in a slot, with the probe across the top of the slot.

## 6.9 HV BUSHINGS

Diagnostic tests on 220 kV, 20 MVA oil impregnated paper (OIP) bushings of a start-up transformer which has been in service for 25 years have been presented in Table 6.12. This data is quite typical of OIP bushings and seem to indicate that the transformer bushings are in order. One of the more important methods of diagnosing bushing degradation is to monitor the dissipation factor and its change with respect to stress and the capacitance. Should there be a measurable and increase in trend in PD measured at the gushing tap and examination of the bushing is warranted. There are reasons to believe that both capacitance and dissipation factor increase with time at constant voltage stress. Furthermore in the absence of PD data, a change (increase) in capacitance of greater than 5% indicates either there is a suspect case of large PD in the vicinity of highly stressed grading foils or a breakdown of the interlayer capacitance. Although the Table 6.12 does not include PD data, it is strongly suggested that PD for bushings be monitored with respect to PD at a later epoch of time.

**Table 6.12.** Typical tests results on a HV bushing

<table>
<tr><th colspan="2" rowspan="2"></th><th rowspan="2">Test connection</th><th colspan="3">Measured values</th></tr>
<tr><th>Test voltage, kV</th><th>Tan δ, (%)</th><th>Capacitance, pF</th></tr>
<tr><td colspan="2" rowspan="9">1. Capacitance and tan δ measurement</td><td rowspan="3">R phase</td><td>2.5</td><td>0.285</td><td>465</td></tr>
<tr><td>5.0</td><td>0.277</td><td>465</td></tr>
<tr><td>10.0</td><td>0.265</td><td>465</td></tr>
<tr><td rowspan="3">Y phase</td><td>2.5</td><td>0.261</td><td>465</td></tr>
<tr><td>5.0</td><td>0.264</td><td>465</td></tr>
<tr><td>10.0</td><td>0.265</td><td>466</td></tr>
<tr><td rowspan="3">B phase</td><td>2.5</td><td>0.462</td><td>458</td></tr>
<tr><td>5.0</td><td>0.463</td><td>459</td></tr>
<tr><td>10.0</td><td>0.453</td><td>459</td></tr>
<tr><td rowspan="2">2. IR test</td><td>15 seconds</td><td rowspan="2">Between HV terminal and Test tap (C point)</td><td colspan="3">5 GΩ</td></tr>
<tr><td>60 seconds</td><td colspan="3">20 GΩ</td></tr>
</table>

Resin impregnated paper bushings are of recent origin, its advent and application to transformers was commenced in early 1990. In view of this long time field data on its degradation and ageing behaviour is not available at the present point of time. With a view to assess the probable elapsed/remaining life of a class of resin impregnated paper (RIP) HV bushings for power transformers, a series of laboratory controlled long duration stress accelerated (E and T) tests were performed. The more important dielectric parameters, (C, tan δ, PD, IR) were continuously monitored over a period of 12200 hours. An inspection time slot of 24 hours was adopted and two of the measured properties mentioned were recorded as shown in Fig. 6.24 for demonstrating the computation.

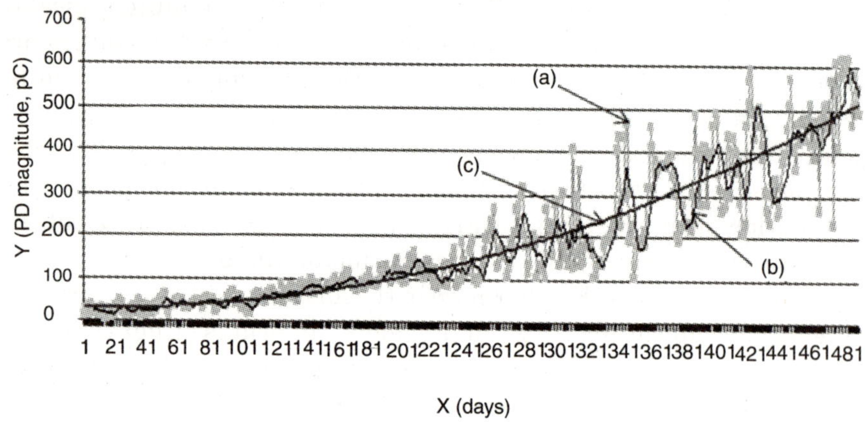


**Fig. 6.24.** Time vs PD magnitude of RIP bushings under accelerated ageing test. (E= 1.25 $E_{normal}$, T= 200°C) **(a)** Measured points, **(b)** Moving average of the data and **(c)** Trend line

The standard deviation of PD magnitude measurement about the trend line was 149 pC and the correlation coefficient with time corresponding to the trend line was 0.85. Equation to the trend line was estimated to be

$$Y = 0.0022\, X^2 - 0.0796\, X + 32.51 \tag{6.23}$$

The further analysis includes a possible estimate of the remaining life, $L$, under normal operating stress conditions ($E_N$ and $T_N$). Suppose, a maximum acceptable limit of PD has been fixed, say 500 pC, it is possible to arrive at a normal operating life based on the earlier studies of the Authors[88]. This procedure is described in the following. For the sake of brevity, only the final expression used in the computation of life is given.

$$L = L_{\text{Accl}} \times \left[\frac{E}{E_N}\right]^n \times e^{B\left(\frac{1}{T_N}-\frac{1}{T_{\text{Accl}}}\right)} \times f(T,E) \tag{6.24}$$

$$L = L_{\text{Accl}} \times K_E \times K_T \times \xi \tag{6.25}$$

In Eqn. (6.25), $K_E$ and $K_T$ are respectively the constants determining the electrical and thermal life and $\xi$ is called the coupling or combined degradation term. The expression for $\xi$ is the following

$$\xi = e^{a\left(\frac{E}{E_N}\right)+\frac{b}{\Delta T}} \tag{6.26}$$

In Eqn. (6.26), $\Delta T$ is the temperature difference between the accelerated temperature (200°C) and the operating temperature based the insulation clause. The epoxy system under consideration is Clause *F* with a maximum operating temperature of 155°C so that $\Delta T$ works out to 45°C. The constants *a* and *b* are difficult to arrive at and intensive experimental runs are required to establish them.

Based on the premise that the electrical ageing has partial discharge as its precursor and temperature has a finite influence on the ageing of the polymeric insulation with paper as a component, Eqns. (6.24 to 6.26) represent the equation to the nominal life under normal operating conditions. It is to be noted that the genesis of these equations is predominantly empirical and the constants involved in the computation determined experimentally, are given in Table 6.13.

**Table 6.13.** The constants of Life Law (Eqns. (6.24 to 6.26))

| | | | | | |
|---|---|---|---|---|---|
| $E/E_N$ | = | 1.25 | $K_E$ | = | 2.73 |
| $\Delta T$ | = | 45.0 | $K_T$ | = | 2.23 |
| B | = | 3600 | $F(T, E) = \xi$ | = | 4.00 |
| n | = | 4.5 | $L_{Accl}$ | = | 11500 h* |
| a | = | 0.4 | L | = | 32 years |
| b | = | 40 | *Calculated for 500 pC from Eqn. (6.23 ) | | |

The temporal changes of the dissipation factor were monitored in an exactly similar way as described earlier. This data has been included in Fig. 6.25. Estimation of remaining life can be made using similar equations, however, the values of the constants therein differ significantly.

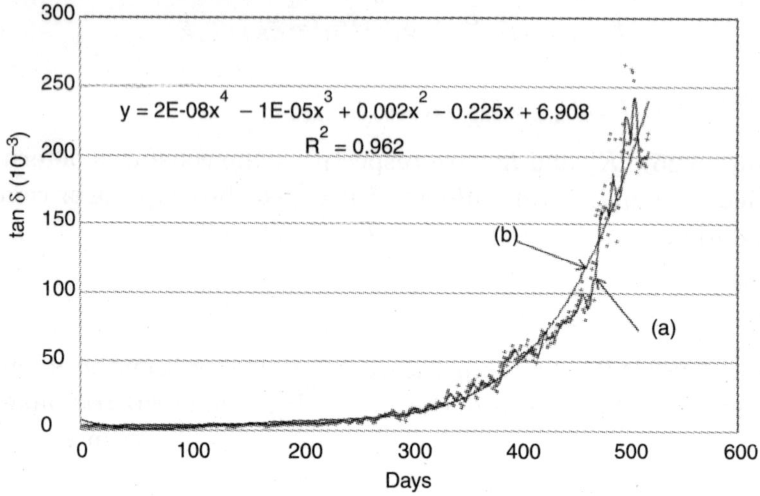


**Fig. 6.25.** Time vs. tan δ of RIP bushings under accelerated ageing test. ($E$ = 1.25 $E_{normal}$, $T$ = 200 ° C) (a) Moving average of the data, (b) Trend line

## 6.10 SUMMARY OF CONVENTIONAL ELECTRICAL DIAGNOSIS

In summary, high voltage insulation diagnostics as is known at the present point of time is a complex concept and involves monitoring many properties thereof. The comprehensive database so accrued is to be processed using refined statistical techniques. It often happens that the test results may give conflicting information on the status of insulation. In such an event, the results are to be checked against earlier records and repeat some of or all the tests. Final analysis and decision making in pronouncing the condition of insulation is almost always guided by sound experience.

The present practice is to develop software-oriented "expert systems" appropriately trained to decipher the data and establish meaningful functional relationships between several parameters entering into the diagnostic schedule. Many expert systems are available for condition monitoring of different power equipment especially for power transformers and have been installed in power stations. This new addition has undoubtedly contributed to improved reliability of the system as a whole and prevented catastrophic failure of expensive station equipment.

As has been mentioned earlier, the DTCM models being data-driven, its relevance, sufficiency and faithfulness, is of great importance. In a practical condition monitoring situation, the data base is obtained from electricity receiving (utility) or generating station records. Assuming that the diagnostic measurements are made carefully and data are recorded meticulously, additional information on the equipment loading patterns including the extent and duration of overloads, information on previous tests, both the invasive and non-invasive, servicing and overhaul schedules should also be made available. This input enhances the reliability of monitoring programs and helps in taking appropriate decisions on the tests to be carried out and the inspection time slots.

## 6.11 EQUIPMENT SPECIFIC ACOUSTIC PD TEST

Acoustic PD test has been carried out on some of the equipment for which electrical PD measurement was not practical. High voltage power capacitors, GIS, power transformers and shunt reactors are considered for acoustic PD described as under.

### 6.11.1 Testing of Power Capacitors for Acoustic PD

Measurement of PD in power capacitors pose serious practical problems, due to non-availability of PD free voltage source of very high rating of the order of 6800 kVA, 50 kV for testing capacitors of rating 50 μF, 20 kV. Also, detection and measuring devices cannot handle such high currents. As a result of this, PD test on power capacitors is not included in the testing standards. However, during laboratory testing of power capacitors, particularly during long term tests (for example over voltage cyclic test and ageing test as per IEC 60871-2 [94]) of duration 1000 or 3000 hours, the need for periodic monitoring of PD to assess degradation of capacitor insulation has been felt by some of the capacitor manufacturers, which will serve as an input for improvement in design, material and construction. In view of this, acoustic PD was measured on capacitors in the laboratory, details of which are furnished in the subsequent sections.

#### *6.11.1.1 Acoustic PD Measured Prior to Breakdown of Capacitor*

A HT capacitor of rating 1000 kvar, 14.4 kV PP film shunt capacitor was tested for finding out its breakdown strength and acoustic PD was monitored. Figure 6.26 shows the arrangement for acoustic test on the capacitor. As depicted in Figs. 6.27 (a) and (b), there was a gradual increase in acoustic PD magnitude with applied voltage from 30 dB at 0.8 $U_n$, to 100 dB at 2.0 $U_n$, where the capacitor failed.

Fig. 6.26. Acoustic PD test on a 1000 kvar, 14.4 kV shunt capacitor

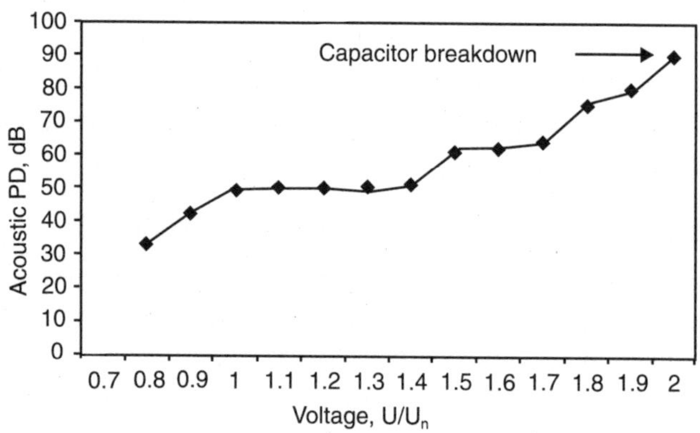


(a) PD amplitude (max) vs voltage

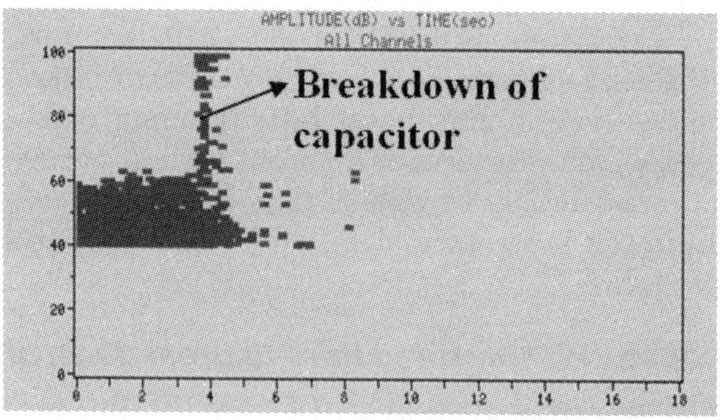


(b) PD amplitude vs time

Fig. 6.27. Acoustic PD signals recorded prior to breakdown of a capacitor

### 6.11.1.2 Acoustic PD Test during Over Voltage Cycle Test

Over voltage cyclic (OVC) test as per IEC 60871-2 [94], is one of the important tests to ascertain the performance of power capacitors during repeated over voltage applications, which simulates repeated switching transients on to the capacitors. At present, no techniques are available to assess the performance of capacitors during this test. To accomplish the same, acoustic PD technique, which is on-line and non-invasive was employed without disturbing the capacitors under test[61, 62].

Two capacitors (oil impregnated PP film type), each of rating 181 kvar, 6.9 kV, designated as CAPH #1 and CAPH #2 were tested for OVC. Table 6.14 gives the summary of their performance along with the acoustic PD monitored at different epochs of OVC test. Acoustic PD monitored during application of a few OVC is shown in Fig. 6.28. One of the capacitors (CAPH # 1) which was showing higher dB level from the beginning of the test, failed during 3rd day of test. The other capacitor, CAPH #2, successfully withstood the test and acoustic PD magnitude remained almost stable during OVC test. Hence, acoustic PD test can be employed in the laboratory for assessing the performance of capacitors during some of the critical tests.

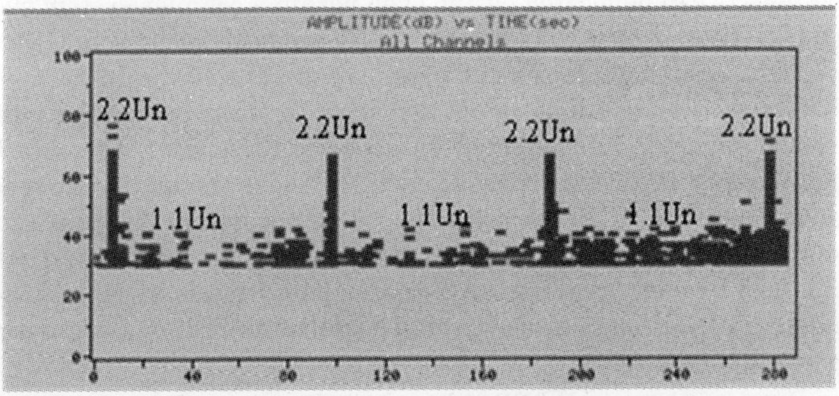


**Fig. 6.28.** Acoustic PD monitored during application of a few cycles of OVC

**Table 6.14.** Acoustic PD signals during OVC test on HV capacitors

| Acoustic PD measurement | OVC test condition | Duration of acoustic PD measurement | Magnitude of Acoustic PD, dB max | |
|---|---|---|---|---|
| | | | Capacitor CAPH#1 | Capacitor CAPH#2 |
| Before OVC test | 6.9 kV (Un ac) | 5 minutes | 70 | 60 |
| $1^{st}$ day of OVC | 15.5 kV (2.25 Un ac) | 10 cycles | 90 | 80 |
| $3^{rd}$ day of OVC (from $400^{th}$ to $410^{th}$ cycle)[(i)] | 15.5 kV (2.25 Un ac) | 10 cycles | >100(ii) | 85 |
| $5^{th}$ day of OVC test (from $740^{th}$ to $750^{th}$ cycle) | 15.5 kV (2.25 Un ac) | 10 cycles | Could not be measured as capacitor failed during $418^{th}$ cycle | 85 |
| After OVC test | 6.9 kV (Un ac) | 5 minutes | | 62 |

Note: (i) Capacitor CAPH#1 failed few cycles after this measurement

(ii)100 dB is the maximum measurable magnitude of acoustic signal.

## 6.11.2 Acoustic PD Measurement on GIS

Acoustic PD detection is simple and cost effective method, which can be applied without built-in sensors, and the short detection range ensures that defects can be located easily, even within a complex system such as a GIS. Locating the PD source is of particular importance in GIS, as this helps in both diagnosis and prognosis of PD activity. In an added advantage, the technique is not sensitive to electro-magnetic interference; however it is affected by acoustic noise. But, this can be minimized by restricting measurement to above the audio-frequency range. Since the AEPD is non-invasive in nature simpler enclosure mounted instrumentation like acoustic sensors, and pre and post amplifiers are only required. A further application is in successively detecting a loose but non-discharging part using vibrations originating from acoustic emission.

Acoustic detection of PD is based on the retrieval and analysis of mechanical signals caused by pressure waves set up due to any of the forms of electrical discharges for instance, PD in a dielectric surrounding a conductor, or corona. The pressure waves propagate through the fluid medium and interact with the walls of the equipment (GIS, Transformer, Power capacitors etc.) and setup small amplitude vibrations. The signals thus generated can be detected by appropriate acoustic sensors.

The following are the various types of defects causing PD are;

i. Moving particles impacting on the enclosure or on an insulating surface

ii. Electrode protrusions resulting in corona

iii. Fixed particles on an insulating surface, also causing corona

iv. Floating electrodes causing a very large-energy PD

v. Voids in solid insulation or de-lamination

A schematic arrangement for AE PD measurement is shown in the Fig. 6.29. Modern spectrum analyzer has a built-in digitizer so that an external digitizing facility is not required. The pre-amplifier is a tuned amplifier to correspond with the frequency range of the sensors. Narrow band sensors with operating frequency between 35 – 100 kHz with a resonant frequency of about 90 kHz and wide band sensors with a frequency range of 40 – 400 kHz with a resonant frequency of about 150 kHz are normally used in the detection. The position of the sensors is to be carefully adjusted such that the attenuation of the signal is the minimum. It is a good practice to use a number of sensors located over a large surface area of the test object.

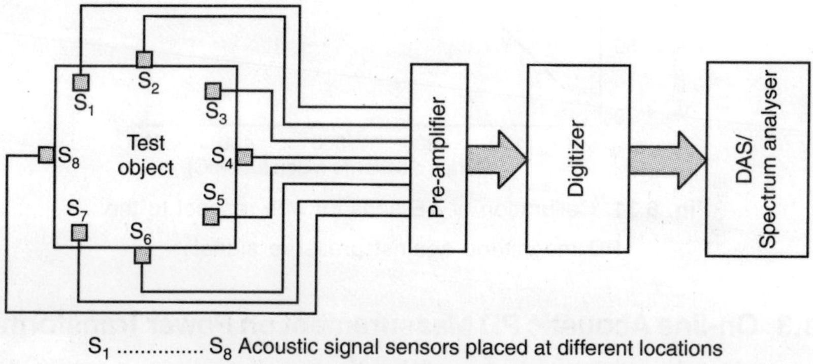


**Fig. 6.29.** A schematic arrangement for measuring acoustic emission signals from a GIS module

The sensors register different parameters responsible for PD aided vibration of the enclosure walls of the test object. For example, the velocity and the pressure signal magnitude (proportional to the acceleration of the enclosure wall at the location of each sensor) can be recorded. In order to obtain the magnitude of the discharges in electrical terms (in pC), a previous calibration of the sensors under controlled laboratory conditions is to be made. Figure 6.30 and Fig. 6.31 give some examples of such calibration.

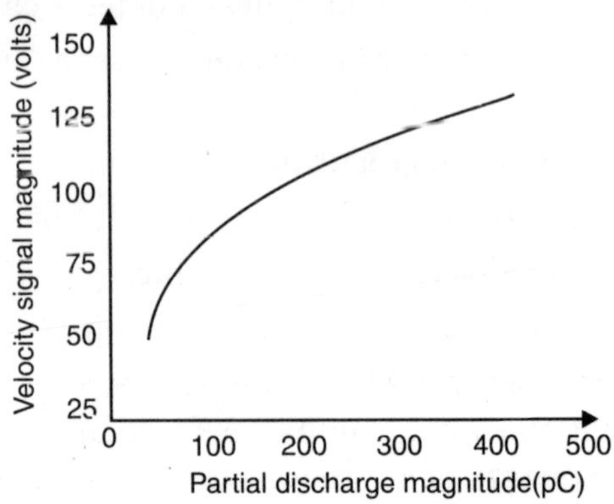


**Fig. 6.30.** Calibration of AE sensors with respect to the PD magnitude against velocity signal

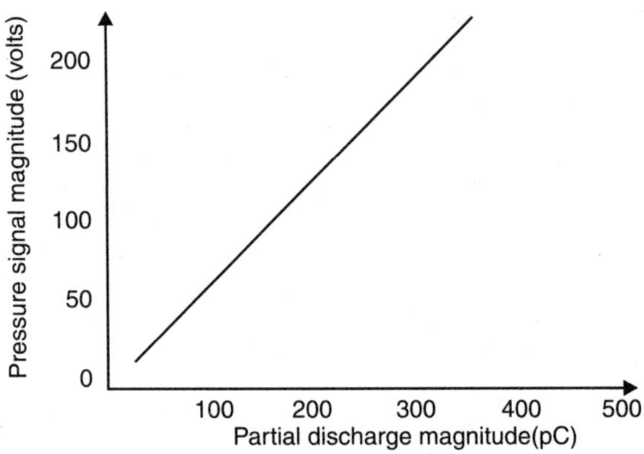


**Fig. 6.31.** Calibration of AE sensors with respect to the PD magnitude against pressure signal

## 6.11.3 On-line Acoustic PD Measurement on Power Transformers

AE technique is employed as one of the diagnostic tools for on-site testing of oil filled transformers, shunt and neutral reactors. More than 150 equipment have been tested by the Authors' laboratory as on July 2008 [48,65]. Typical ratings of the power transformers tested cover a wide range from 50 MVA to 315 MVA and voltage class of 400/13.8 kV, 220/110kV, 220/66 kV, 220/33 kV and 110/33 kV. Transformers tested are of 5 to 35 years old.

### *6.11.3.1 Mounting of AE Sensors*

AE Sensors are mounted on the tank walls of the transformer using magnetic holders, after cleaning the surface and applying silicone or couplant grease for proper sensor contact. Identification numbers are assigned to the sensors and their coordinates (x,y,z) are noted with reference to a convenient origin (0,0,0). The sensor mounting positions vary with the size, design and rating of the transformer and test conditions. Figure 6.32 shows an example of sensor mounting layout. For single phase units with square sides, four sensors are mounted on each face of the transformer tank wall. It is useful to maintain the same sensor positions for transformers of similar design for comparison purposes.

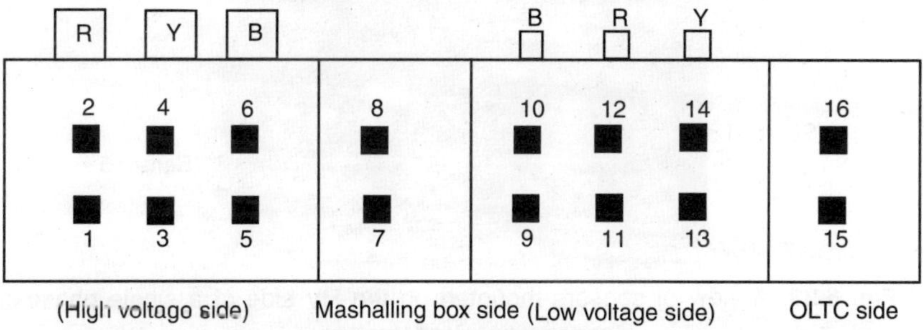


**Fig. 6.32.** Mounting scheme for sensors on a three phase transformer

Necessary safety precautions are to be taken during mounting the sensors which is often done under energized conditions of the transformer. It is recommended not to approach the high voltage bushing flange of the transformer. Movement of ladder or other climbing devices should be handled with extreme precaution. It is not necessary to access the top of the transformer for AE testing.

Figures 6.33 and 6.34, respectively, show mounting of sensors on

(i) 43.33 MVA, $220/\sqrt{3}$ kV, single phase generating station unit transformer for acoustic PD measurement at site, and

(ii) 250 MVA, 15.75/400 kV, three phase generator transformer.

### *6.11.3.2 Instrument Setup at Site*

Sensors mounted on the transformer are connected to the AE system through shielded co-axial cables and checked for proper connections. The software settings like threshold, minimum frequency, maximum frequency, sampling rate etc., are properly selected. The test is carried out generally for 1-15 minutes. Before performing AE testing, it is recommended to collect logistic information regarding the transformer history, results of other diagnostic tests and maintenance undertaken.

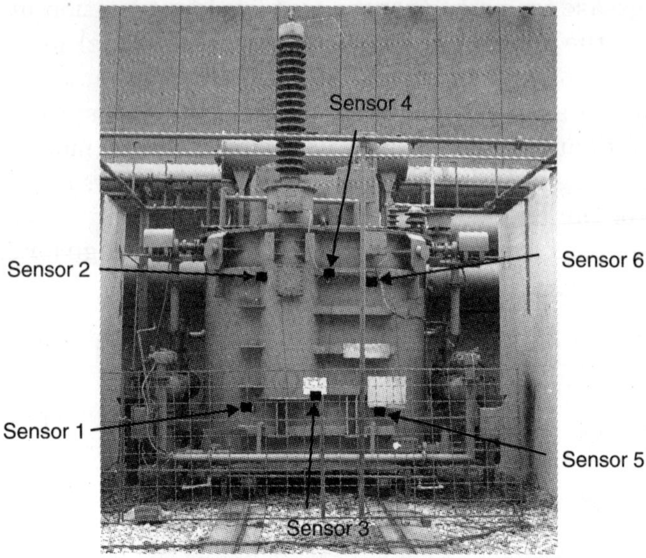


**Fig. 6.33.** A view of sensors mounted on the HV side of a single phase generating transformer

During the test necessary information like load conditions, tap changing operations, pump operations, if carried out are noted down which will be of great help for data interpretation and analysis.

#### 6.11.3.3 Case Study on Oil Filled Power Transformers of Rating up to 400 kV

Acoustic PD testing was performed on power transformers of rating up to 400 kV at various sites of power utilities and state electricity boards. Typical ratings of transformers tested range from 50 MVA to 315 MVA and voltage class of 400 /13.8 kV, 220 kV/110 kV, 220 kV/66 kV, 220 kV/33 kV, 110 kV/33 kV, and so on. Power transformers tested include both single phase and three phase units. Figure 6.35(a) shows eight Generating Station Unit (GSU) transformers of rating 11 kV/ (220 /√3) kV, 43.33 MVA, 1-phase being tested for acoustic PD at one of the hydro power stations. Figure 6.35(b) depicts the Acoustic PD magnitude for one of the GSU transformers.

More than 180 transformers have been tested by the Authors, for acoustic PD. Table 6.15 gives the summary of on-line testing of power transformers and Fig. 6.36 shows the magnitudes of acoustic signals obtained from different transformers of rating up to 400 kV at various substations. The highest acoustic PD amplitude measured so far is 88 dB for a 100 MVA, 230/110 kV transformer which was tripping frequently and showing fluctuating DGA results. Transformers with acoustic PD below 60 dB have been reported to be working satisfactorily.

**Table 6.15.** Summary of acoustic PD test on transformers [48, 65]

| Voltage rating | Number of transformers tested | Measured value of acoustic PD, dB |
|---|---|---|
| < 220kV | 25 | 35 - 79 |
| 220kV | 68 | 35 - 88 |
| 400kV | 59 | 35 - 80 |

Sensor 1 and Sensor 3 are not visible

**Fig. 6.34**. A view of sensors fixed on the HV side of a 3ϕ generator transformer

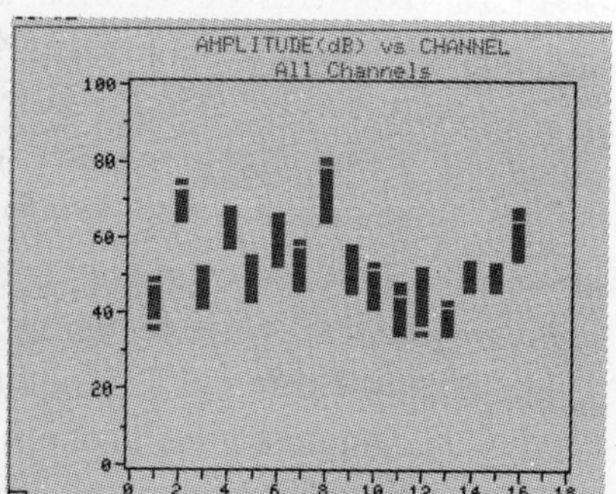


(a) (b)

**Fig. 6.35.** Acoustic PD test on 220kV 43.33MVA transformers in a hydrogenating station. **(a)** A view transformers, **(b)** Acoustic PD magnitudo

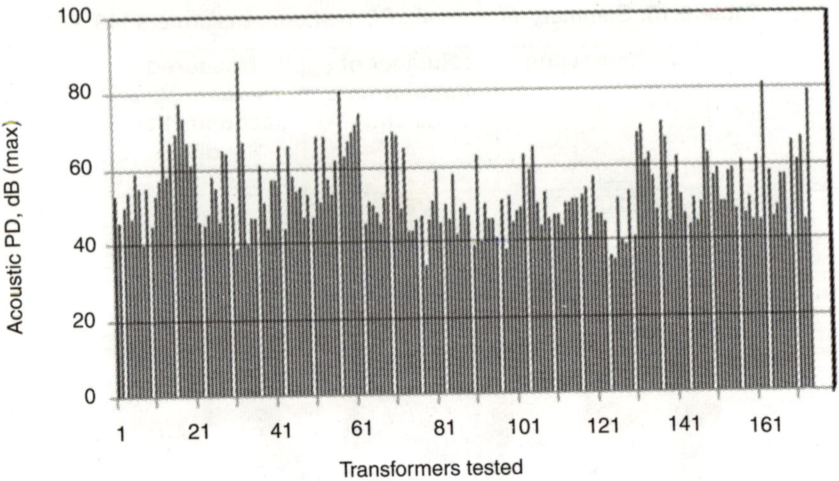


**Fig. 6.36.** Acoustic PD magnitude for power transformers of rating up to 400 kV

### *6.11.3.4 Transformer Diagnosis Employing Conventional and Non-conventional Techniques*

For assessing the insulation condition of power transformers at site, several electrical diagnostic tests are performed. As electrical PD test at site on EHV grade transformers was not practically feasible, non-conventional acoustic PD method was employed. Other diagnostic tests were off-line conventional tests, namely, Polarization Index (PI), Capacitance and tan δ, DC leakage current, Residual Voltage Measurement (RVM) and FRA. Two typical case studies are compared for a healthy transformer and a defective transformer,

**Case- 1:** Results of on-site diagnostic tests on a 33 years old transformer of rating 220 kV/33 kV, 20 MVA are summarized in Table 6.16. Measured values indicate low PD activity and satisfactory results of other diagnostic tests. The transformer insulation appears to be healthy.

**Case- 2:** Results of on-site diagnostic tests on a 10 years old transformer of rating 220kV /33 kV, 35 MVA are summarized in Table 6.17. Measured values indicate high PD activity and poor results of other diagnostic tests like high tan δ value, deterioration of oil. Transformer insulation, therefore appears to be unhealthy.

These case studies indicate among other things the efficacy of using acoustic sensing method in discriminating healthy and faulty equipment.

The AE test results included in Tables 6.16 and 6.17 indicate reasonably clearly that, at least on a qualitative basis it is possible to discern the status of insulation in transformers. There is some debate on the application of AE technique to quantitatively estimate the amount of degradation in electrical insulation. The reason for this apprehension stems from the fact that more often the energy associated with PD is deemed to be more relevant in assessing the condition of the transformer and often this information is not forth coming when AE technique is applied.

In summary, it can be reported that with advent of new and revised techniques of acoustic emission measurement and quantification, it seems possible to use AE PD as a strong index of insulation degradation. Particularly in very large equipment where electrical methods have been known to be precluded.

**Table 6.16.** On-site diagnostic test results on a 220 kV, 20 MVA transformer

| Test | | Measured values at test connection | | | Remarks |
|---|---|---|---|---|---|
| | | (i) | (ii) | (iii) | |
| Acoustic PD at Un, dB | | 47 | | | Acoustic PD activity is low |
| Insulation Resistance at 5 kV dc, MΩ | 60 sec | 1050 | 1020 | 1010 | Test results are satisfactory and therefore, transformer insulation appears to be healthy. |
| | 600 sec | 2150 | 2050 | 1230 | |
| Polarization index | | 2.05 | 2.01 | 2.05 | |
| Capacitance, nF at 2.5 kV | | 21 | 22 | 27 | |
| Tan delta (%) at 2.5 kV | | 0.5 | 0.5 | 0.5 | |
| DC leakage Current, μA at 8 kV dc | | 30 | 30 | 30 | |
| RVM test at 2 kV rms | | Smooth polarization spectra, without dominant peak, indicating homogeneous condition of transformer insulation. Estimated moisture content in transformer insulation is below 2 %. | | | |
| FRA | | Results are normal. No winding deformation / displacement. Mechanical integrity is healthy. | | | |

(i) HV winding versus LV winding shorted to grounded tank,
(ii) HV winding versus LV winding ungrounded,
(iii) LV winding versus HV winding shorted to grounded tank

**Table 6.17.** On-site diagnosis of a 220 kV, 35 MVA transformer

| Test | | Measured values at test connection (i) | (ii) | (iii) | Remarks |
|---|---|---|---|---|---|
| Acoustic PD at Un, dB | | 85 | | | High PD activity |
| Insulation Resistance at 5 kV dc, MΩ | 60 sec | 1050 | 1020 | 1010 | Tan δ values are high, bdv of oil is low and therefore, transformer insulation appears to be not healthy. |
| | 600 sec | 2150 | 2050 | 1230 | |
| Polarization index | | 2.05 | 2.01 | 2.05 | |
| Capacitance, nF at 2.5 kV | | 19 | 20 | 20 | |
| Tan δ (%) at 2.5 kV | | 8.2 | 9.4 | 7.5 | |
| DC leakage Current, μA at 8 kV dc | | 200 | 210 | 300 | |
| BDV of oil | | < 30 kV | | | |
| DGA result | | Fluctuating DGA values | | | |

### *6.11.3.5 Periodic Acoustic Testing of Transformers*

Periodic on-line condition monitoring of a number of transformers, which include power transformers, NGT, Rectifier transformers at various locations and sub stations has been performed. For the transformers showing higher PD activity (more than 60 dB), necessary maintenance works were suggested. A typical case of periodic testing of a few transformers is summarized in Table 6.18, wherein acoustic PD magnitude recorded during the year 2002, 2004 and 2006 are shown along with maintenance work carried out are shown. As can be seen from Table 6.18, periodic AE testing is found to be effective in bringing down the PD activity to safe levels.

**Table 6.18.** Periodic acoustic PD testing of transformers

| Transformer details | Magnitude (dB) of acoustic PD measured during years 2002 | 2004 | 2006 | Maintenance performed |
|---|---|---|---|---|
| Transformer-1, 25 MVA, 33/6.6kV | 68 | 66 | 56 | Oil filtration, Tightening of terminals, Megger value improved from 24 to 58Meg Ohms |
| Transformer-2, 5.8MVA, 33/0.8kV | 71 | 64 | 59 | Oil filtration, Cleaning, Tightening of terminals. |
| Transformer-3, 5.8MVA, 33/0.8kV | 74 | 73 | 60 | Oil filtration, Cleaning, Tightening of terminals, plugging of leakage |
| Transformer-4, 5.8MVA, 33/0.8kV | 75 | 74 | 62 | Oil filtration, Tightening of terminals, plugging of leakage |

### 6.11.4 Acoustic PD Measurement on Shunt Reactor

A shunt reactor of rating 63 MVAR, 420 kV was reported to show high gas content and fluctuating DGA results. Due to some practical issues, it was not possible to carryout conventional electrical PD test at site. On-line acoustic PD test (refer to Fig. 6.37) showed AEPD signals of magnitude 67 dB. The 3D location chart (refer Fig. 6.38) showed concentrated AE activity in the area where the neutral wire was taken out of the reactor for NGT connection. Post test analysis confirmed loose connection of the neutral wire, leading to arcing near that region.

**Fig. 6.37.** Acoustic testing of a shunt reactor

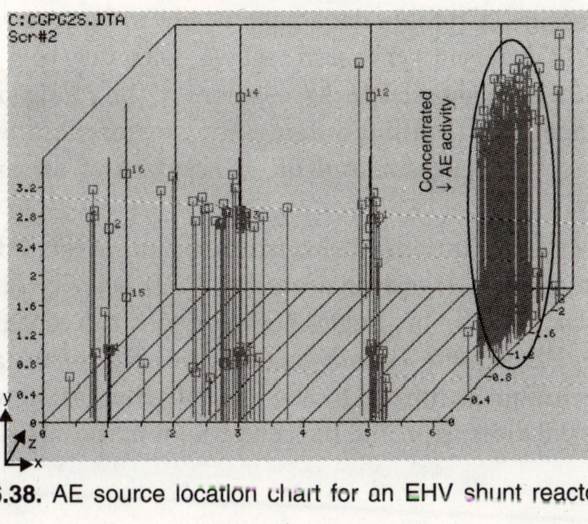


**Fig. 6.38.** AE source location chart for an EHV shunt reactor

## 6.12 SUMMARY OF ACOUSTIC PD TEST

Acoustic technique is non-destructive, non-invasive, on-line and the equipment to be tested need not be removed from the circuit. Therefore, power shutdown is not required for conducting on-line acoustic PD test at site, which is an attractive aspect for substation/power station personnel. Acoustic emission technique has been customized for detection and location of partial discharge and other types of defects in power equipment, particularly, transformers which will be of great help to the power utilities to identify the defective transformers and weeding out the faulty ones before they fail during service leading to catastrophic consequences. Custom built multi-channel acoustic systems (up to 52 channels) are available for testing power equipment of rating up to 400 kV. Most of the commercially available acoustic systems are compact, portable and can be easily set up at site. Defects of several kinds can be localized with an accuracy of 10% on laboratory scale specimens. However, for location of defects in power equipment at site more inputs like material, design, construction and type of transformer are required from the manufacturers and utilities.

Acoustic emission technique provides information useful for diagnosis and also helps to schedule operation and maintenance of power transformers. Signature analysis based on waveform and FFT is used for discrimination of AE PD signals from noise.

Elimination of noise enhances the accuracy of source location. Acoustic Emission technique reduces guess-work in equipment diagnosis and maintenance cost by localization of fault and the condition at which this fault occurs. Credible data base needs to be generated for power equipment of different types and ratings, both new and aged, to arrive at safe limits for acoustic PD. Assessment of the condition of insulation based on acoustic technique demands considerable experience and expertise of the operator and often services of an expert becomes mandatory.

One of the limitations of acoustic technique is related to attenuation of acoustic signals. However severe the PD activity or advanced the detection system may be, unless signals reach the container and hence the sensor, PD activity cannot be detected. Therefore, applicability of acoustic technique is limited by the level of attenuation which is related to the physical dimension and the constituent materials of the test object.

# APPENDIX - 1

# Possible Defects in Power Equipment

| Electrical equipment | Defects |
|---|---|
| XLPE power cables, terminations and joints | i. Voids, imperfections, impurities, moisture, defective raw materials.<br>ii. Non-uniformity in insulation thickness (eccentricity).<br>iii. Sharp projections in semi-conducting(SC) layers towards insulation, improper contact between SC and insulation surface.<br>iv. Inadequate skill employed while making joints and terminations.<br>v. Inadequate technique to prevent ingress of moisture during manufacturing, transportation, storage, installation or in service.<br>vi. Improper protective coverings for cables terminations and joints when they are installed in polluted and fire hazardous environment. |
| Oil impregnated paper cables | i. Voids, imperfections, impurities, moisture, defective raw materials.<br>ii. Improper stress control and inadequate skill employed while making joints and terminations.<br>iii. Pressure leak.<br>iv. Oil leak.<br>v. Inadequate moisture barriers (metallic cover) to prevent ingress of moisture during manufacturing, transportation, storage, installation or in service.<br>vi. Improper protective coverings for cables terminations and joints when they are installed in polluted and fire hazardous environment. |
| Power capacitors | i. Air voids in the PP film during processing, Defective polypropylene film.<br>ii. Non uniformity of insulation thickness.<br>iii. Defective metallization – poor quality and composition of metal or alloy used for metallization of polypropylene film capacitor. |

*Contd...*

| | |
|---|---|
| | iv. Air voids between the film layers due to incomplete impregnation.<br>v. Inclusion of impurities like dust and other foreign particles in between the film layers.<br>vi. Stress concentration at the electrode edge.<br>vii. Swelling of PP film, which interrupts the flow of the fluid along any micro-gap between adjacent films leading to air voids.<br>viii. Gas evolution from impregnant under high voltage stress.<br>ix. Poor workmanship like improper winding. |
| Gas insulated systems | i. Moving particles impacting on the enclosure or on an insulating surface.<br>ii. Electrode protrusions resulting in corona.<br>iii. Fixed particles on an insulating surface, also causing corona.<br>iv. Floating electrodes causing a very large-energy PD.<br>v. Voids in solid insulation. |
| Power transformers | i. Defective insulating materials, voids, imperfections, impurities, moisture in insulation<br>ii. Improper bushing lead connection.<br>iii. Stress concentration in oil leading to gassing of transformer oil.<br>iv. Poor workmanship.<br>v. Inadequate technique to prevent ingress of moisture during manufacturing, transportation, storage, installation or in service. |
| HV machines | i. Inadequate impregnation or Dipping.<br>ii. Loose stator coils.<br>iii. Semi-conductive coating failure.<br>iv. Scarf joint failure.<br>v. Electrical tracking.<br>vi. Inadequate enwinding spacing.<br>vii. Thermal deterioration / cycling. |
| HV bushings | i. Voids, imperfections, impurities, moisture, defective material.<br>ii. Loose conductor lead connection.<br>iii. Inadequate technique to prevent ingress of moisture.<br>iv. Leakage of oil leading to dryness in insulation.<br>v. Improper stress grading. |

APPENDIX - 2

# List of IEC Standards Covering PD Measurement on Different Electrical Equipment

| | Equipment | IEC standard covering PD measurement | Title |
|---|---|---|---|
| 1 | HV Electrical Equipment : High-voltage test techniques – PD measurement | IEC 60270: 2000 | Partial discharge measurements. |
| 2 | Rotating electrical machines | IEC 60034-18-42: 2008 | Qualification and acceptance tests for partial discharge resistant electrical insulation systems (Type II) used in rotating electrical machines fed from voltage converters. |
| | | IEC 60034-27: 2006 | Off-line partial discharge measurements on the stator winding insulation of rotating electrical machines. |
| 3 | Power transformers | IEC 60076-3: 2000 | Insulation levels, dielectric tests and external clearances in air. |
| 4 | Power cables with extruded insulation | IEC 60502-2: 2005 | Power cables with extruded insulation and their accessories for rated voltages from 1 kV (Um = 1,2 kV) up to 30 kV (Um = 36 kV) – Part 2: Cables for rated voltages from 6 kV (Um = 7,2 kV) up to 30 kV (Um = 36 kV). |
| | | IEC 60840-2004 | Power cables with extruded insulation rated voltages above 30 kV up to 150 kV. |

*Contd...*

| | | | |
|---|---|---|---|
| | | IEC 62067:2000 | Power cables with extruded insulation and their accessories for rated voltages above 150 kV (Um = 170 kV) up to 500 kV (Um = 550 kV) – Test methods and requirements. |
| | | IEC 62067-am1:2006 | Amendment 1 – Power cables with extruded insulation and their accessories for rated voltages above 150 kV (Um = 170 kV) up to 500 kV (Um = 550 kV) – Test methods and requirements. |
| | | IEC 60885-2: 1987 | Electrical test methods for electric cables. Part 2: Partial discharge tests. |
| | | IEC 60885-3: 1988 | Electrical test methods for electric cables. Part 3: Test methods for partial discharge measurements on lengths of extruded power cables. |
| 5 | HV Circuit Breakers | IEC 62271-100: 2001 | HV Switchgear and Control gear: AC Circuit breakers. |
| 6 | Power Capacitors | No IEC standard | PD test not covered. |
| 7 | HV Current Transformers | IEC 60044-1: 2003 | Instrument Transformers – Current Transformers. |
| 8 | Connectors for electronic equipment | IEC 60512-4-2: 2002 | Tests and measurements - Part 4-2: Voltage stress tests - Test 4b: Partial discharge. |
| 9 | Insulating liquids | IEC/TR 61294:1993 | Determination of the partial discharge inception voltage (PDIV) - Test procedure. |
| 10 | Electrical insulating materials and system | IEC 61934: 2006 | Electrical measurement of partial discharges (PD) under short rise time and repetitive voltage impulses. |

# APPENDIX - 3

## Acoustic Testing Terminology

1. Amplitude distribution : A display of the number of AE signals of Amplitude ≥ A pre-set level.
2. Attenuation : Loss of amplitude with distance as the wave travels through the structure.
3. Burst emission : A qualitative description of the discrete individual emission event.
4. Channel (AE) : A single AE sensor and the related components for transmitting, conditioning, detecting and measuring the signals those come from it.
5. Continuous emission : A qualitative description of the sustained signal level due to a rapidly occurring acoustic emission event.
6. Counts : The number of times the AE signal crosses the detection threshold ("ring down counts", "threshold crossing counts").
7. dB (deci Bel) : A unit of measurement for AE signal amplitude, $A$, $A = 20 \log Vp$ ($Vp$ is the peak signal magnitude (mV) referred to the preamplifier input).
8. Detection (AE) : Recognition of the presence of signal (accomplished by signal crossing the threshold).
9. Duration : The elapsed time between the first threshold crossing to the last.
10. Event (AE) : A local material change giving rise to acoustic emission.
11. Event data set (AE) : The set of numbers describing an event pursuant to data processing that recognizes a single event producing more than one hit.

12. Event description : A digital description of an event, comprising one or more signal information extracted from them or calculated from them.
13. Event energy : The total elastic wave energy associated with an acoustic emission event.
14. Felicity effect : (The reverse of the Kaiser effect defined later): The presence of AE at stress levels below the maximum previously experienced.
15. Guard sensors : A sensor intended to eliminate of noise based on arrival times.
16. Hit data set : The set of numbers representing random signal features
17. Intensity (AE) : A measure of the size of the emission signals detected, such as the average amplitude, average AE energy or average counts.
18. Kaiser effect : The absence of detectable acoustic emission at a fixed level until previously applied stress levels is exceeded.
19. Noise : Signals produced by causes other than AE, or by AE sources that are not relevant to the purpose of the test.
20. Nondestructive diagnostic testing : A technical term describing a non-invasive, or, a minimally invasive test procedure for detecting incipient or slowly developing faults in an equipment. As the name suggests, the test object is not subjected to any stress which might endanger its intended operability either during or at a later/time as a consequence of the test.
21. Rise-time : The time taken by an AE signal to attain the maximum value from its first threshold crossing.
22. Sensor (AE) : A device, called a transducer that converts an acoustic signal to an electrical (voltage) signal.
23. Signal (AE) : The electrical signal output of the transducer realized after appropriate amplification and frequency filtering.
24. Signal features : Measurable characteristics of the AE signal, amplitude, energy, duration, counts, rise-time, that can be stored in a memory device
25. Source energy (AE) : The total energy dissipated by the AE process in the transducer.

# References


1. Zalewskai J. and Skubis J., "Evaluation of Partial Discharges by the Acoustic Emission Method", CIGRE Symposium, Paper 700-02, Vienna, 1987.
2. Lundgaard et al., "Ultrasonic Detection of Particle Movement and Discharges in GIS", Gaseous Dielectrics, Vol. 3, Pergamon Press, NY, 1982.
3. Kreuger F. H., "Discharge Detection in High Voltage Equipment", Temple Press, London/ American Elsevier, NY, 1964.
4. Ramu T. S. and Reddy C. H., "Reliability and Life Estimation of Power Equipment" (Book), New Age International Publishers, India, 2009.
5. Whitehead S., "Dielectric Breakdown of Solids" (Book), Clarendon Press, Oxford, 1951.
6. Rogers E. C. and Skipper D. J., "Gaseous Discharge Phenomena in High Voltage DC Cable Dielectrics", Proc. Inst. Elect. Ingg., p.107, 1960.
7. Kreuger F. H., "Partial Discharge Detection in High-Voltage Equipment" (Book), Butterworths Publication, UK, 1989.
8. Densley J. and Salvage B., "IEEE Transactions" EI Vol. 6, pp. 54–62, 1971.
9. Badran I.N.S., El-Assi, Riaz O.M.B., "Savage B.", Proceedings Int. ISH, Munich, 1972, pp. 396–402.
10. Lemke E., Ein Beitrag Zur, Mssung Impulsloser Teilent-ladungen bei wechselspannung, Elektrie Vol. 30, pp. 479–482, 1976.
11. Kerner H.C., "Degradation Parameters for Insulating Materials and Diagnostic Criteria Applicable also to Equipment", CIGRWE WG 15.06, 1989
12. Konig D., "Impulslose Teilentladungen in hohlraumen von epoxydharz – formstoff - isolierungen", ETZ-A, Vol. 19, pp. 156–158, 1969.
13. Ramu T.S. and Smit J.J., "Measurement and Characterization of Micro Discharges (Pulse-less discharges) in Electrical Insulation", KEMA Research Report 82021-EIO 90-303, R&D Division, KEMA, Arnhen, The Netherlands, 1990.
14. Kodoll W.R., Karner H. C., Tanaka T., Idoa M., "Internal Partial Resistivity Testing", CIGRE No.15.04, 1988.

15. Pradeep Kumar Shetty, "Recognition and Retrieval of Partial Discharges in Equipment – A Statistical Signal Modelling and Feature Extraction Approach", Ph.D. Thesis, Indian Institute of Science, Bangalore, India, 2004.
16. Tanaka T., Okamoto T., "An Advanced Partial Discharge Measurement Method Sensitive to the Shape of a Void", CIGRE 1988, Paper No. 15-02.
17. Srinivas M.B. and Ramu T. S., "Pulse-count Distribution as a Possible Diagnostic Tool for Assessing the Level of Degradation of Rotating Machine Insulation," IEEE Trans. on Electrical Insulation, USA, Vol. EI-25, Aug. 1990, pp. 747-756.
18. Muhr M., Strehl T., Gulski E., Feser K., Gockenbach E., Hauschild W., Lemke E., "Sensors and Sensing used for Non-conventional PD Detection" D1-102, CIGRE Working Group D1 102, CIGRE 2006.
19. Mallikarjunappa K., "High Frequency Current Transformer Technique for Detection of Partial Discharges in Power Capacitors", M.Sc. (Engg.) Thesis, Indian Institute of Science, Bangalore, India, July 1993.
20. Bartnikas R. & McMohan E. J., "Engineering Dielectrics, Vol. 1 – Corona Measurements and Interpretation" (Book), ASTM Special Publication, STP 669, Year 1979.
21. Kubler B., "Inegrierende Brucke Mit Optoelektronischer Messwertuberstragung fur Teilentladungsutersuchungen", ETZ, Vol. 98, pp. 237–239.
22. Dakin T.W., Malinaric P.J., "A Capacitance Bridge Method for Measuring Integrated Corona Charge Transfer Per Cycle", Trans AIEE Vol. 79, pp. 468–453, 1960.
23. Miller R. K., McIntire P., Acoustic Emission Testing, Vol. 5, 2nd Edition, Nondestructive Testing Handbook, Ed., American Society for Nondestructive Testing, 1987, pp. 167–186, 187–193.
24. Parry D. L., "Industrial Application of Acoustic Emission Analysis Technology in Monitoring Structural Integrity by Acoustic Emission", STP 571, American Society for Testing and Materials, 1975, pp. 194–202.
25. Miller R. K. and McIntire P., Acoustic Emission Testing, Vol. 5, 2nd Edition, Nondestructive Testing Handbook, Ed., American Society for Nondestructive Testing, 1987, pp. 58–61, 84-88.
26. Pollock A. A., Acoustic Emission Inspection, Physical Acoustics Corporation, TR 103-96-12/89.
27. Lundgaard L.E., Partial discharge -Part- XIII: Acoustic Partial Discharge Detection – Fundamental Considerations IEEE Electrical Insulation Magazine July/August 1992- Vol. 8, No. 4. pp. 25–31.
28. Bartnikas R. & Mc Mohan E. J., Engineering Dielectrics, Vol. 1 – Corona Measurements and Interpretation, ASTM Special Publication, 1979, p. 327.

29. Tord Bengtsson et al., Partial Discharge Studies using Acoustic Emission, Contribution to NORD.IS 94, Vaasa, Finland, June 13-14, 1994.

30. Ogihara H., EEJ9, pp. 12–21, 1964.

31. Harrold R. T., Corona Measurement and Interpretation, Chapter 10, Vol. 1, Edited by Bartnikas and McMahon ASTM, STP 669, 1979.

32. Howells E., and Norton E. T., Detection of Partial Discharges in Transformers using Acoustic Emission Technique, Paper No. 77 585-3, presented at IEEE Summer Power Meeting, Institute of Electrical and Electronics Engineers, July 1977.

33. Harrold R.T., Acoustic Wave-guides for Sensing and Locating Electrical Discharges within High Power Transformers and other Apparatus, IEEE Winter Power Meeting, Institute of Electrical and Electronics Engineers, New York, 1978.

34. Kawada H., et al., Partial Discharge Automatic Monitor for Oil Filled Power Transformers, IEEE Transactions on Power Apparatus and Systems, Vol. PAS 103, No. 2, February 1984.

35. EPRI EL – 4009 Project 426 – 1 Report: Acoustic Emission Detection of Partial Discharges in Power Transformers.

36. James R.E., Phung B.T., Development of Computer-based Measurements and their Application to PD Pattern Analysis, IEEE Transactions on Dielectrics and Electrical Insulation, Vol. 2, No. 5, 1995, pp. 838–854.

37. Stone G.C., Partial discharge Part-VII: Practical Techniques for Measuring PD in Operating Equipment. IEEE Electrical Insulation Magazine, Vol.7; No.4; July/August, 1991; 9-19.

38. Lundgaard L. E., Partial Discharge-Part XIII: Acoustic Partial Discharge Detection-Fundamental Considerations, IEEE Electrical Insulation Magazine July/August 1992 – Vol. 8, No. 4.

39. Lundgaard L. E., Partial Discharge–Part XIV: Acoustic Partial Discharge Detection –Practical Application, IEEE Electrical Insulation Magazine Sep./Oct. 1992 Vol. 8, No. 5.

40. Peter M. E., Physical Acoustics Corporation, Partial Discharge – Part XXI: Acoustic Emission-Based PD Source Location in Transformers, IEEE Electrical Insulation Magazine, 1995.

41. Lee Y. S., Nelson J. K., Scarton H. A., Teng D. and Azizi S., An Acoustic Diagnostic Technique for use with Electric Machine Insulation" 1070-9878/94/ IEEE, 1994.

42. Harrold R.T., Member IEEE Westinghouse Research and Development Center Pittsburgh, Pennsylvania 15235, Acoustic wave guides for sensing and locating Electrical Discharges in High Voltage Power Transformers and other Apparatus, 0018-9510/79/0300-0449, IEEE, 1979.

43. Ron Harrold, Westinghouse Science and Technology Center, Partial Discharge – Part XVI: Ultrasonic Sensing of PD within Large Capacitors, IEEE Electrical Insulation Magazine, May/June 1993, Vol. 9, No. 3.

44. Bozzo R., Guastavino F., Guerra G., PD Detection and Localization by Means of Acoustic Measurements on Hydogenerator Stator Bars, 1070-9878/1995 IEEE.
45. Reynders J.P., Landy C.F, Meyer C.F, Wolmarans D. W., Experience in the Use of Electrical and Acoustic Diagnostic Procedures on a Significant Population of Similar Large High Voltage Motors, Year 1992 Session, 30 August – 5 September, 3-5, re to Metz 75010 Paris (CIGRE).
46. Nandini Gupta and Ramu T.S., Indian Institute of Science, Bangalore, Estimation of Partial Discharge Parameters in GIS using Acoustic Emission Techniques: a theoretical approach, Journal of Computational Acoustics 6, 1998, pp. 403–419.
47. Sarathi R., Vishal Dubey and Srinivasa Y.G., Identification of Incipient Discharges in Gas Insulated System using Acoustic Emission Technique - Workshop on AE Technique for Condition Monitoring of Power Transformers Employing AE Technique, 10-11 February 2005, CPRI, Bangalore, India.
48. Nagamani H.N., Bhavani Shanker T., Vaidhyanathan V. and Neelakantan S., Acoustic Emission Technique for Detection and Location of Simulated Defects in Power Transformers - presented in the IEEE Power Tech 2005 conference, St. Petersburg, Russia, 27–30 June 2005.
49. Beldi F., The Brown Boveri Review, Vol. 37, No. 6, June 1950, pp. 179–193.
50. Anderson, J.G., Transactions on Power Apparatus and Systems, American Institute of Electrical Engineers, Vol. 75, December 1956, pp. 12–21.
51. Harrold R.T., Acoustic Wave Guide for Sensing and Locating Discharges within High Voltage Power Transformers and Other Apparatus, IEEE Winter Power Meeting, 1978.
52. Paul – Emile Vohl et al., Model Analysis of Ultrasonic Pulse Generated by Partial Discharges, IEEE-PAS, May 4, 1976.
53. Channakeshava, Some Aspects of Acoustic Detection in Transformers, Conference and Workshop on EHV Technology, Bangalore, August 20–23, 1984.
54. David Train, Georger, H., Vallana Court, Ryszard Melewski, Comparison of Three Detection Techniques of Partial Discharge Measurements in Power Transformers, IEEE–PAS, Vol. 104, 1985, pp. 900–909.
55. Muhr M., IEC 62478 – A Prospective Standard for Acoustic and Electromagnetic Partial Discharge Measurements, HIGHVOLT KOLLOQUIUM' 07, Dresden, Germany, 14–15 May, 2007.
56. Kristiansen H.O. and Lien K., Acoustic Diagnosis of Gas insulated substation– International Conference on Condition Monitoring and Diagnostic Engine0anagement of Power Station/Substation Equipment – CPRI, Bangalore, India, 2009.

57. Markalous S.M. and Bolliger A., Case Studies Using Acoustic Emission and UHF Probes to Detect and Locate Partial Discharges in Power Transformer – International Conference on Condition Monitoring and Diagnostic Engineering Management of Power Station/Substation Equipment –CPRI, Bangalore, India, 2009.
58. Claude Kane and Alexander Golubev, On-line Transformer Diagnosis, International Conference on Condition Monitoring and Diagnostic Engineering Management of Power Station/Substation Equipment – CPRI, Bangalore, India, 2009.
59. Tang Liang, Wang Weilling, Acoustic Location of Partial Discharges in On-line Power Transformers, Sixth ISH, New Orleans L.A., USA, August 29–September 1, 1989, p.22.
60. Blackburn T. R. et al., An Improved Electric/Acoustic Method for the Location of Partial Discharges in Power Transformers.
61. Nagamani H.N., Channakeshava, AE Technique to Monitor the Performance of HV Capacitors during Endurance Test as per IEC 60871-2:1999, presented at the 2004 Doble Clients Conference, March 21–26, 2004, Westin Copley Place, Boston, Mass, USA.
62. Nagamani H.N., Bhavani Shanker T., AE Technique to Monitor the Performance of HV Capacitors During Laboratory Testing – presented at the 5th International Workshop on Acoustic Emission – at SHAR Center, Shriharikota, November 2002.
63. Nagamani H.N., Bhavani Shanker T., Moorching S.N., CPRI, India, Detection and Location of Partial Discharges in Power Capacitors using Acoustic Technique, Sixth-International Seminar on Electrical and Electronic Insulating Materials-INSULEC–2000, 23–24 November 2000, Hyderabad, India.
64. Nagamani H.N., A Study on Detection and Location of PD in Power Capacitors Employing Acoustic Emission Technique - Thesis Ph.D., 2004.
65. Nagamani H.N., Bhavani Shanker T., Proceedings of the Workshop on, On-line Diagnosis of Power Transformers by Acoustic Emission Technique, 10–11 February, 2005, CPRI, Bangalore, India.
66. Lundgaard L.E., Partial Discharges – Part XIV: Ultrasonic Sensing of PD within Large Capacitors, IEEE Electrical Insulation Magazine May/ June 1993, Vol. 9, No.3.
67. ASTM E 1316–02: Standard Terminology for Non-destructive Examinations.
68. Insulation Levels, Dielectric Tests and External Clearance in Air for Power Transformers. IEC 60076–3, 2000.
69. Nagamani, H. N, Bhavani Shanker T., Detection and Location (3D) of Multiple PD Sources Employing Acoustic Emission Detection Technique, presented at INSULEC-2005, the 7th International Seminar on Electrical & Electronic Insulating Materials & Systems, 20–21 Jan 2005, Mumbai, India.

70. IEEE Std C57.127-2000 – IEEE Trial-use Guide for the Detection of Acoustic Emissions from Partial Discharges in Oil Immersed Power Transformers.
71. Arturo Nunez, Samuel J. Ternowchek, R K Miller, Barry Ward, EPRI, Detection, Location, Assessment & Location of Faults in Energized Power Transformers using Acoustic Emission, CIGRE Transformer Colloquium 2003, Merida, Yucatan, Mexico.
72. DISP User's Manual of M/s. Physical Acoustics Corporation, USA, 2001.
73. Acoustic Emission Sensors, Product Bulletin, Physical Acoustic Corporation, 1997.
74. Technical Write-up on Noesis for AE data Analysis by M/s. Physical Acoustics Corporation, USA, 2003.
75. Mallat S., A Wavelet Tour of Signal Processing (Book), Academic Press, Chapter 7, 1999.
76. Robert H. Shumway and David S. Staffer, Time Series Analysis and its Applications (Book), Second Edition, Springer Publication, 2005.
77. Hamming R.W., Digital Filters (Book), Dover Publication, NY, 1998.
78. Book by Ramu T.S., Diagnostic Testing and Life Estimation of Power Equipment, Publisher – New Age Science, Control Center, 11A, Little Mount Sion, Turnbridge, Wells, Kent, TN1, 1YS, UK, International Edition, 2009.
79. Gulski E., Wester F. J., Schikarsl, Insulation Condition Assessment of MV Power Cables using PD Detection and Analysis at Oscillating Voltages, ERA Technology, Surrey, Leatherhead, 2000.
80. Virsberg L.G., Kelen A., Some Observations on the Very Low Frequency Testing of HV Machine Insulation, CIGRE, Paper No. 108, 1964.
81. Miller R., Black I.A., Partial Discharge Measurement Over the Frequency Range 0.1Hz to 50 Hz, IEEE Trans. On EI, Vol. 12, No. 2, pp. 2240–233, 1977.
82. Eager G.S., Katz C., Fryszczyn B., Densely J., Bernstein B.S., High Voltage VLF Testing of Cables, IEEE Trans on Power Delivery, Vol. 12, No. 2, pp. 565–570, April 1997.
83. Asano, K., Anno, K., Higashiyama, Y., The Behaviour of Charged Conducting Particles in Electric Fields, IEEE Transactions on Industry Applications, Volume 33, Issue 3, pp. 679–686 May–June 1997.
84. LeNy, R., The Electrostatic Problem of a Portion of a Cylinder Protruding from a Plane Electrode in an Electric Field, Journal of Electrostatics, Vol. 11, p. 97, 1981.
85. Meijer S., Gulski E., Smit J. J., Wester F. J., Grun T., Turner M., Interpretation of PD in GIS using Spectral Analysis, International Symposium on High Voltage Engineering (ISH), Vol. 5, 1999, Paper No. 5.12.516, London, UK.

86. Mole, G., Design and Performance of a Portable AC Discharge Detector, ERA Report, V/T 115, 1952.

87. Mason, J. H., The Breakdown of Insulation by Electrical Discharges, Proceedings IEE, Vol. 100, Part II A, pp. 149–164, 1954.

88. Ramu, T. S., On the Estimation of Life of Power Apparatus Insulation Under Combined Electrical and Thermal Stresses, IEEE Tr. EI, Vol. 20, No. 1, pp. 70–78, February 1985.

89. Hemalatha B. and Ramu T.S., Insulation Degradation under Multifactor Stress, Fifth International Symposium on High Voltage Engineering, Braunschweig, West Germany, August 1987.

90. Ramu, T.S., Degradation of HV Generator Insulation under Mechanical Electrical and Thermal Stresses, Invited Lecture, IEEE Symposium, Electrical Insulation, Toronto, Canada, June 1990.

91. Wichmann, A., Grunewald, P., Influence of Dielectric Stress Concentration on the Voltage Endurance of Epoxy Mica Generator Insulation, IEEE Tr. EI, Vol. 12, pp. 428–434, 1977.

92. Kelen, A., Functional Testing of High Voltage Generator Stator Insulation, CIGRE, Paper No. 15.03, Vol. 1, 1976.

93. Dakin W., Works C.N., Thomson J.S., Electromagnetic Probe for Detecting and Locating Discharges in Large Rotating Machine Stator, IEEE Trans. On EI, Vol. 88, pp. 251–257, March 1969.

94. IEC 60871-2,1999: Shunt Capacitors for AC Power System having a Rated Voltage above 1000V- Part2- Endurance Testing.